Carolina F.M. Santos
Jéssica K.S. Pachú
José B. Malaquias

Insecticidal plants and agro-homeopathy

Carolina F.M. Santos
Jéssica K.S. Pachú
José B. Malaquias

Insecticidal plants and agro-homeopathy

tools for pest management

ScienciaScripts

Imprint

Any brand names and product names mentioned in this book are subject to trademark, brand or patent protection and are trademarks or registered trademarks of their respective holders. The use of brand names, product names, common names, trade names, product descriptions etc. even without a particular marking in this work is in no way to be construed to mean that such names may be regarded as unrestricted in respect of trademark and brand protection legislation and could thus be used by anyone.

Cover image: www.ingimage.com

This book is a translation from the original published under ISBN 978-620-3-03889-7.

Publisher:
Sciencia Scripts
is a trademark of
International Book Market Service Ltd., member of OmniScriptum Publishing Group
17 Meldrum Street, Beau Bassin 71504, Mauritius
Printed at: see last page
ISBN: 978-620-3-52119-1

INSECTICIDAL PLANTS AND AGRO-HOMEOPATHY: TOOLS FOR PEST MANAGEMENT

Carolina Francine Mariano Santos [1]
Jéssica Karina da Silva da Silva Pachú [2]
José Bruno Malaquias [3]

Mayerli Tatiana Borbón Cortés

(Translator)

[1]Researcher at UNESP.
[2]Postdoctoral researcher in Biostatistics at IBB-UNESP, malaquias.josebruno@gmail.com;
[3] D. student in Entomology at the University of São Paulo, ESALQ campus jessikapachu@gmail.com.

1

PRESENTATION

Pest control through the use of pesticides has been subject to several problems, due to the risk of resistance selection in insects, negative influence on populations of non-target organisms and environmental contamination. In this context, other practices are necessary, such as biological control and the use of alternative insecticides in order to minimize the negative impacts of chemical contamination. In this sense, homeopathic preparations, such as those obtained from dried seeds of the plant *Delphinium staphisagria* L. (Ranunculaceae), which contain toxic alkaloids, have received relevant attention from the scientific community and producers, especially with respect to aphid control. However, it is important to present information that can support the implementation of this molecule within an ecologically based pest management program. In view of this scenario, the present publication aims to report the potential of agro-Homeopathy for ecological pest control and its potential effects on natural enemies.

INDEX

CONSIDERATIONS ON HOMEOPATHY

Brazil has stood out over the years as one of the countries with the highest agricultural production and consequently the demand for synthetic insecticides used to control insect pests has also increased. On the other hand, consumer preference for products without chemical residues (synthetic insecticide residues) has been increasing. In this context, the conventional production system is losing ground to alternative production systems. With the implementation of alternative production systems, the risks of pollution and intoxication of farmers and consumers are reduced, making organic and/or alternative agriculture a possibility to reduce the use of synthetic insecticides in agricultural production.

With the advent of modernization in agriculture, agroecosystems underwent a sequence of disturbances marked by the adoption of monoculture practices, such as the use of external inputs such as chemical fertilizers and pesticides, as well as high mechanization. Despite having boosted productivity, these practices have triggered a series of problems that influence ecological dynamics and favor phytophagous species, which often find in monoculture a large availability of food resources and few or no natural enemies capable of controlling the population. These conditions lead to an increase in the population of these phytophagous species and contribute to their characterization as pests (SUJII *et al.*, 2010).

Pest control exclusively through the use of pesticides loses its effectiveness with the risk of resistance selection and reduction of natural enemy populations, as well as the other socio-environmental impacts derived from the application of chemical pesticides (GLIESSMAN, 2007). In this context, alternative management strategies become very necessary. Alternative pest management can be an extremely important practice for scenarios of this nature, given its evident potential for pest containment through the use of predators, parasitoids or pathogens, which has proven to be quite successful.

The use of natural enemies in agroecosystems does not represent a novelty in pest management and even preceded the use of xenobiotics (PRIMAVESI, 2016). There is evidence of the use of predatory ants *Oecophylla smaragdina* Fabricius (Hymenoptera: Formicidae) to suppress lepidopteran and coleopteran populations in the 3rd century BC in China. in China. In 1888, California imported from Australia the ladybug *Rodolia cardinalis* Mulsant (Coleoptera: Coccinellidae) to combat infestations of *Icerya purchasi* Maskell (Hemiptera: Monophlebidae) in citrus, being this case considered as the first successful program of classical biological control (PRIMAVESI, 2016).

Biological control is a phytosanitary control tool that can be associated with other methods and applications, thus integrating a set of technologies known as Integrated Pest Management (IPM). These actions seek to consider effective and at the same time ecologically viable and sustainable measures to control species that cause damage to agricultural crops, based on the application of more than one method concomitantly, aiming to guarantee the productivity of the system with the minimum possible impact (PRIMAVESI, 2016).

The use of botanical insecticides as an alternative or complementary method for pest management was documented by Isman (2006), who has indicated their use as an insecticide or repellent. However, there is a significant demand for studies that are able to experimentally demonstrate the efficacy of these products, especially in the context of pest management in Brazilian agriculture. Currently, the studies conducted have indicated greater use in the context of insects of medical and veterinary importance (WYNN & FOLGERE, 2006), being nonexistent systematic studies on the use of plants in the broader context, such as agriculture. Agro-homeopathics with insecticidal effect are products with high potential for use as tools to boost the sustainability of agroecological systems in tropical climate countries (ROSSI *et al.*, 2007a). The extraction of homeopathic products for this purpose has been carried out from plants, among which *Delphinium staphisagria*, a species popularly known as albarraz or staphisagria, belonging to the Ranunculaceae family, stands out. The product is obtained from the dried seeds of this plant, which contains toxic alkaloids. *Delphinium staphisagria* was used in the formulation of homeopathic preparations with insecticidal effect documented by Davidson (1929), who noted its potential in the control of aphids, among other arthropods.

Alternative production systems are mainly characterized by not using synthetic insecticides and chemical fertilizers, instead they use technologies that respect ecological principles, prioritizing the preservation of natural spaces and the conservation of biodiversity. In these systems, as well as in others, the presence of pests and diseases in crops is very common, being necessary the use of natural and/or alternative products, such as homeopathic products (LUCKMANN, 2013).

Homeopathic science was born in 1796 after the publication of the scientific article entitled: "Essay to discover the curative virtues of medicinal substances, followed by some comments on the curative principles admitted up to our days" (FHB, 2011). It is a word of Greek origin meaning "similar disease" (*homoios* = similar and *pathos* = suffering, disease) (ROSSI *et al.,* 2007).

It is a scientific method for the treatment and prevention of acute and chronic diseases, where the cure is given through non-aggressive medicines that stimulate the body's reaction, strengthening its natural defense mechanisms. Therefore, it constitutes an alternative medicinal technique that contemplates the totality of the human being, plants and/or animals to the detriment of isolated diseases. It acts by means of energetic stimuli triggered by homeopathic medicines with the purpose of rebalancing the vital energy of patients.

Homeopathy originated hundreds of years ago, after the philosopher Hippocrates in 16th century Europe. Theophrastus Bombastus Von Hoheinheim, also known as Paracelsus, restored the merits of the treatment of like by like. It was only at the end of the 18th century that this medical specialization was studied in more detail by a German physician, Samuel Hahnemann, which was extremely rare at the time, and yet he himself formalized it into a true system of medicine.

From 1970 and 1980, several studies were carried out by researchers from several countries, such as Germany, Italy, Switzerland, England, Scotland, South Africa, Cuba and Mexico. In Brazil, the first works were developed at the Federal University of Viçosa, under the supervision of Professor Dr. Vicente Wagner Dias Casali, starting in 1998.

Conceptually, Barbosa (2006) proposes that the homeopathic physician seeks, in addition to physical alterations of the patient, other signs and symptoms that characterize that individual as a whole so that from the parts (symptoms) the whole (the patient) is reached; this reasoning seeks homeopathic procedures that not only treat the disease, but mainly the patient and with that it is possible to find the most similar medicines to achieve the cure of the treated being.

According to Hamly's (1982) perspective, Homeopathy is a comprehensive conception of disease, which allows the homeopathic physician to treat his patients on the expectation of being able to cure many common and supposedly incurable processes, in a less dangerous way than that used by the conventional schools.

According to Reinhart (1993), in Homeopathy there are still no studies that confirm side effects as evidenced in almost all treatments with allopathic medicines. Since homeopathy does not present such negative effects, it can be applied for certain homeopathic purposes. The science of Homeopathy is based on observation, experimentation and the recognition and respect for the laws of life. The principles of Homeopathy apply to any level of complexity (SOUZA & RESENDE, 2011).

According to Araújo (1999), Homeopathy restores the environment without residues, favoring the natural development of insects and microorganisms, which are responsible for the transformation and incorporation of organic matter in the soil. Thus, Homeopathy contemplates the balance of plants, which will be constituted or transformed together with the animals that consume them, resulting in food with higher quality for human consumption.

According to Rezende (2003), to be a homeopath of plants and livestock or soil animals implies having knowledge, conscience, respect and ethics in action. It means respecting the eternity of vital processes, whatever the belief or religious inclination of his fellow man and the people who live in his community. It is necessary that the veterinarian makes the work of awareness of how to use the treatments with Homeopathy and to make a greater follow-up to the patient.

According to Schembri (1992), Homeopathy is the science that treats sick individuals (and not diseases) through medicines prepared in infinitesimal dilutions capable of producing symptoms in healthy individuals similar to the symptoms confirmed in the sick, that is, it treats the disease with medicines that cause a disease.

Homeopathy with its knowledge and resources is indicated as the tool for the transformation of sick living systems into healthy, balanced and sustainable living systems. However, the work of helping the Kingdoms of nature can be assumed by everyone and everywhere, since it starts in the way we relate to life (SOUZA & RESENDE, 2011).

The plant kingdom constitutes the major source for the preparation of homeopathic medicines, the plant can be used in the various vegetative phases, complete and/or its parts, such as: subway part, leaf, flower, bark, wood, rhizome, fruit and seed; of which are also used products of their extraction or transformation: juice, resin, essence, etc. (FHB, 2011).

Biotherapeutics or nosodes are chemically undefined products that serve as raw material for the dynamized preparations, which can be: secretions, pathological or non-pathological excretions, products of microbial origin, among others. This technique of using dynamized preparations is known as Isopathy (*iso* = equal and *pathos* = suffering, disease). Similarly, homeopathic solutions can be prepared with plants in order to balance their development in the culture environment (ROSSI *et al.,* 2007a).

Generally, applications in agriculture are based on the principle of isopathy, since it does not include the collection of unhealthy signs, necessary for the homeopathic practice itself. However, the increasing knowledge of plant physiology allows describing symptoms and physiological responses in plants, with certain similarity to those observed in humans, and therefore, it is possible to think about the selection of medicines based on symptomatic similarity (BONATO, 2007). The application of homeopathy in agriculture is a technology of very low cost and easy application (ROSSI *et al.,* 2007a). It is a science that can be applied to all living beings, aiming at their equilibrium (ROSSI *et al.,* 2007a).

Among the environmental conditions that cause damage to plants are drought, flooding, high or low temperature, salinity, mineral deficiency in the soil, excess or lack of light and also phytotoxic compounds such as O3 (ozone), which can cause damage to plant tissues (BONATO, 2007). The author comments that resistance or sensitivity to stress depends on the species, genotype and developmental age of the plants.

Murray (1998) addresses another issue, the interaction between the animal's internal biological environment and management, and its influence on healthiness. He also comments that the animal's immune system, in order to resist infection, is affected by diet, stress, exercise, presence of other infections or damage, external environmental conditions (vigor and competitiveness of existing beneficial natural microorganisms) and the genetics of the animals.

In general, stress triggers a broad response in plants, ranging from altered gene expression and cellular metabolism to altered growth rate and productivity. Thus, plant responses to stress depend on the duration, severity, number of exposures and combination of stressors, as well as on the type of organ and tissue, developmental age and genotype (BONATO, 2007).

Bonato (2009) proposes that biotic and abiotic factors cause imbalance in the vital energy, and that when somatized it results in disease for the plant or in a minimal physiological disorder, but such disorder can lead the plant to death or reduce its productivity, depending obviously on the biological plasticity of the plant species studied. However, when a homeopathic medicine capable of producing the same symptoms in the plant is applied, the result will be the restoration or minimization of the harmful effects caused by biotic and abiotic factors on the vital energy.

Biotic factors such as pests and diseases also cause stress in plants, one of the most important in agriculture being those caused by phytopathogenic organisms. However, plants possess mechanisms that, depending on the virulence of the pathogen, can prevent or reduce the damage caused by phytopathogens, and one of these mechanisms is called resistance induction (BONATO, 2007).

As a consequence of stress, plants may show resistance or susceptibility, which may result in survival or death, respectively. But what is the relationship between the types of stress described so far and the homeopathic medicines themselves? As described below, when homeopathic medicines are applied in a rational and timely manner, and mainly obeying the Law of Similars, the survival increases and the death of plants decreases. First of all, it is important to comment that for homeopathic science any disorder caused in the plant, either by biotic or abiotic factors, first affects the vital energy (vital principle, vital force) of the plant. Thus, whenever the plant is subjected to a certain stress, it is rigorously with its vital energy unbalanced and consequently dissonant of its natural homeostasis (BONATO, 2009).

The Law of Similars consists in the use of a medicinal substance that has the ability to produce, in healthy people, artificial symptoms similar to the symptoms of the patient who wants to be cured. The homeopathic medicine should make the organism activate its natural defenses against disorders, acting on the recovery of the vital energy of the individual (BITENCOURT & BONATO, 2008).

It is essential to take care of the patient in addition to the disease because it is a fundamental piece. And so, by treating the patient feels better and consequently will recover faster and return to their daily life or in the case of plants, their environment will improve for it to reproduce and grow with quality.

According to Barbosa (2006), the conventional physician seeks to collect some data from the patient that indicate a cause among the countless possible ones. Starting from the whole (the patient), one arrives at the part (the disease). On the other hand, Homeopathy starts from the patient to the disease and only then seeks to apply the appropriate treatments.

Homeopathy helps traditional medicine and must be performed, but alone it does not recover the being as with homeopathic treatments, and that is why it is often necessary to use allotropic medicines so that together they achieve the total cure of the patient who undergoes this treatment.

Homeopathy is often indicated for anxiety/depression problems and chronic diseases such as: diabetes, hypertension, arthritis, vices, allergies, tendonitis, gastritis, colitis, obesity, vasculitis, psoriasis, lupus, thyroiditis, nephritis and several other chronic diseases.

Araújo's vision defends that the environment is produced in balance with the other treatments practiced with Homeopathy, but it is necessary to be careful when applying these treatments since the environment needs all the beings that inhabit it and by exterminating any of them, the environment as a whole may be at risk.

The vegetable or animal production, making use of the techniques of Homeopathy, will achieve more profit, both financially and environmentally and as a producer in search of this balance must produce within the limits that nature and animals offer to not stagnate the production that they provide to humans, and consequently everyone wins with these techniques, which are bringing great benefits to the health of beings and the environment in which they reside.

It is an alternative treatment that is given from the dilution and dynamization of the substance that causes disease to the healthy individual, and as it is a low cost therapy its use is increasing and bringing many benefits to the health of individuals with a particular disease. The therapy seeks to help the patient to achieve a better quality of life, seeking to take care of the patient and not only the disease itself, but the living being that needs care and treatment performed according to the capacity of his body.

Homeopathy seeks to concentrate more on the patient, providing conditions for him to act and prevent the installation of diseases in his organism. As analyzed by the aforementioned author, it is necessary to interact with the entire surrounding environment to analyze the infectious agents causing the disease, trying to treat it preventively.

Finally, it can be said that the use of homeopathy in agriculture is dynamic and shows evidence of good results for animal and plant health, but its use must be accompanied by specialized professionals because its misuse can cause disasters to the environment.

Homeopathy is above all a science that has no owner. Hahnemann left the following message: "If the laws of nature which I proclaim are true, then they can be applied to all living beings". So why can't we use them to treat plants? It is obvious that we can, besides Homeopathy is liberating, it makes the farmer less a slave of the companies, which gives him more economic independence. In addition, homeopathic preparations have a very low cost, which increases the profit of the producer and what is more important, does not contaminate humans, animals, soil or plants, respecting nature.

The so-called Law of Similars, used in agriculture, is exemplified by the case of the soybean larva (*Anticarsia gemmatalis* Hubner) that is contaminated with the *Baculovirus anticarsia* virus. Spraying the larva with the virus in soybean controls the larva itself. This is a practical application of the Law of Similars. Another example would be the use of the flu vaccine, which uses the same virus to fight the flu itself.

Homeopathy can treat humans, animals, plants and soil. In agriculture, it can be used in the control of pests, diseases, improve crop productivity and in the natural defense of plants. Quite interesting works related to Homeopathy in vegetables were published in India. Verma *et al.* (1989) used Lachesis and Chimaphila (C200), aiming to control Tobacco mosaic virus (TMV), which progressively reduces the productivity of infected plants. Homeopathic solutions applied before and after virus incubation reduced 50% of the virus content in leaf discs (BONATO, 2009). It can also be used on soil, with excellent results. We know that soil has life, and therefore is a living organism, and treatment with homeopathic preparations makes it balanced, especially in soils intoxicated with pesticides. It must always be taken into account that the soil is alive. If the soil is unbalanced, how will the plants growing in it be? Logically they will be sick, although the symptoms are not always visible.

Thus, if we treat the soil from homeopathic procedures, within a rational and/or local management, contributing to the improvement of the soil-plant system as a whole, in a greater and better balance, this can be reflected in good crop productivity.

The homeopathic essence says that there are no diseases, but there are sick people. The disease would be a consequence of the vital imbalance of the organism. If Homeopathy were used obeying its basic principles (Law of Similars), it would stimulate the defense systems of living beings. In simpler terms, it would be the following: a well nourished plant within an organic system would be less likely to get sick or be attacked by pests than, for example, a plant growing in soils contaminated with large amounts of pesticides.

In agriculture it is common, and with excellent results, the application of homeopathic preparations, made with the causal agent of the disease or cause of the imbalance. This is what we call Nosode or Biotherapeutic. Nosodes of pests such as bedbugs, larvae, ants, beetles, etc., and fungi such as anthracnose, rust and viruses are widely used.

Rossi *et al.* 2004, comments that it is important to highlight the advantages of Homeopathic experiments on vegetables: the diversity is very large, that is, it is possible to work from perennial crops such as citrus, for example, to very short cycle crops, such as radish, which is ready to harvest after 28 days. It is easy to do research using seeds and seedlings, which, to make an analogy, would be "similar" to research with children. It is possible to work with large populations, as it is not uncommon to find work in protected environments with 5,000 individuals or more. In the field, this number could be even higher, since there are plants that propagate asexually, which is equivalent to saying that it is simple to conduct research with genetically identical "clone" individuals.

According to Câmara 2010 *apud* Teixeira (1998) and Kent (1996), "the use of Homeopathy in agriculture means environmental quality and safety for rural workers and consumers, because one of its characteristics is the use of infinitesimal concentrations of matter".

Scientific studies show that there are two ways to use Homeopathy in agriculture. The first occurs through the application of products based on medicinal plants and other materials, such as minerals for example, with the objective of revitalizing the cultivated plants, rebalancing them and making them sufficiently structured, not only to perform and externalize all their genetic potential, but also to be able to overcome environmental antagonisms, whether climatic, phytopathogenic, nutritional, physiological or any other. The second is used when phytosanitary problems occur, for example, the incidence of insects and diseases (both bacterial and fungal) and the organism itself is used to control it. In this case, the homeopathic product is called "nosode" and acts according to Hahnemann's principle or concept of similarity, as cited above (CÂMARA, 2010).

Rossi *et al.* (2007a), with the objective of evaluating the production of three potato cultivars with the application of nine homeopathic preparations, verified that these influenced the productivity only of the Aracy cultivar, being that the treatment with *Helianthus* CH12 produced the least and was not statistically different from that with 30% alcohol, and both were inferior to all the other treatments.

In the study by Rossi *et al.* (2007a), on the effect of the homeopathic preparation for the control of coquito (*Cyperus rotundus* L.), succession and diversification of species, they found that the initial results indicated a decrease of approximately 35% of the dry mass of the coquito with the lowest doses (50 and 500 g/ha) of the homeopathic product applied.

According to Arenales (2000), homeopathy provides the producer with an increase in profits due to the increase in production and a decrease in expenses. And, according to Reinhart (1993), homeopathy has been successfully used in animal production in several countries and is shown to be an adequate therapy, since it does not leave residues in food.

AGRO-HOMEOPATHY

Based on the use of dynamic ultradilutions and experiments on healthy individuals, Homeopathy is configured as a method of treatment and cure, based on fundamentals defined by Hahnemann. Although it was founded as an alternative to medicine, it is applicable to all living beings, representing a possibility in the phytosanitary management of agroecosystems, as opposed to the conventional agricultural model, which suffers recurrent problems due to its high ecosystemic impact (BETTI *et al*, 2006; CROWDER *et al.*, 2010). Thus, Agro-Homeopathy corresponds to the branch of Homeopathy dedicated to the study of the effects of ultra-diluted and dynamic products of mineral, plant or animal origin on species of agricultural importance, due to its potential as a tool to promote the sustainability of agroecological systems in tropical climate countries (ROSSI *et al.*, 2007).

In Brazil, homeopathic preparations are recognized as an agricultural input through Normative Instruction No. 7, which establishes the norms for national organic production (BRAZIL, 1999). For their effectiveness in the Brazilian context, it is necessary to develop research with agro-homeopathics from the national demand, considering that studies with these products are still scarce and that there is no specific homeopathic Materia Medica for plants, which would represent an important collection of information and instructions regarding their use in different species and purposes (BRUNINI; SAMPAIO, 1993; FONSECA, 2002; TEIXEIRA; CARNEIRO, 2017).

In the mid 1920's, after the international recognition of Biodynamic Agriculture, based on the principles of Homeopathy, the first investigations with agro-homeopathics began to be published (LOOS, 2006). The first study that demonstrated the insecticidal action of Staphisagria with the CH 6 (centesimal hahnnemanian) potency on aphids was documented by Davidson (1929), and the first literature review on the subject was published in 1984, in which the main methodological flaws in the research published up to that time were raised, such as the absence of details of the methods and the lack of statistical analyses that corroborated the results (SCOFIELD, 1984). From that publication, the scientific rigor in research with homeopathics increased, as well as the number of publications in which products of different origins applied on plants affected by phytopathological problems, pests and nutritional deficiencies were evaluated, in addition to healthy individuals, respecting one of the fundamental principles of Homeopathy (TEIXEIRA; CARNEIRO, 2017).

According to the literature review conducted by Teixeira and Carneiro (2017), among the 48 studies evaluated with MIS ≥ 5 (methodological quality index), published between 1979 and 2015, 29 studies obtained significant results compared to the control. Of the documented works, 23 corresponded to wheat trials, 29 were conducted on healthy individuals, 13 on abiotic stressed plants and 6 of them investigated their influence on phytopathological problems caused by fungi, nematodes and viruses (TEIXEIRA; CARNEIRO, 2017).

In addition, the literature reviews published until then aimed to analyze especially publications in Agro-Homeopathy in a general way, effects of agro-homeopathics in healthy individuals and in plants with abiotic stress, leaving a gap regarding its application as a phytosanitary management method, especially with respect to the susceptibility of plants to insect pests (JÄGER *et al*, 2011; MAJEWSKY *et al*, 2009; TEIXEIRA; CARNEIRO, 2017). In this context, the need for research to be executed with agro-homeopathics on plants of national agricultural importance is evident, in addition to the need for works that aim to fill the gap regarding the insecticidal potential of homeopathic preparations, as well as their effects on non-target organisms. Although work on the influence of agro-homeopathics on insects is still incipient, there are already studies with satisfactory results, which open avenues and possibilities for research in this area.

According to Brunini and Arenales (1993), the agro-homeopathic Staphisagria is capable of satisfactorily helping to contain aphids on vegetables, improving their resistance to the attack of these insects.

In the research conducted by Mapeli *et al.* (2004), resistant and susceptible cabbages, with and without attack of the aphid *Brevicoryne* brassicae (Hemiptera: Aphididae) were used for the preparation of agro-homeopathic preparations of CH 5 dynamization: isotherapy preparations (nosodes made with the damaging agents themselves) produced with aphids with CH 5 and CH 30 dynamization, as well as controls with water without dynamization and 70% alcohol + CH 5 water. The preparation made from the resistant cultivar CH 5 showed a reduction in aphid reproductive rates, as well as a decrease in the immigration rate of winged individuals.

Wyss *et al.* (2010) evaluated the use of the preparation *Lycopodium clavatum* CH 15 and the nosode of the ashy apple aphid *Dysaphis plantaginea* (Hemiptera: Aphididae) CH 6, which enabled the reduction of the number of descendants and the consequent population decrease of the colony.

Fazolin *et al.* (2000) in a study carried out with nosodes on the leafhopper *Cerotoma tingomarianus* (Coleoptera: Chrysomelidae), an important pest of the bean crop, reported that in plants treated with the product, leaf consumption by individuals of this species was reduced, resulting in the death of the insects caused by starvation.

According to Almeida (2003), the agro-homeopathic prepared from *Spodoptera frugiperda* (Lepidoptera: Noctuidae) larvae, dynamized in CH 30 reduced the population of the corn earworm on corn plants with four leaves.

Giesel *et al.* (2007) observed a reduction in the foraging rate of leafcutter ants *Acromyrmex* spp. subjected to treatment with nosodes from individuals of the same genus in CH 30 dynamization. In addition, they conducted trials with *Belladonna* CH 30 and macerated fungi from ant mounds in the same dynamization, which did not show results as effective as nosodes.

Rupp *et al.* (2012) found that in peach trees attacked by fruit flies *Anastrepha fraterculus* (Diptera: Tephritidae), the agro-homeopathic Staphisagria together with nosodes of the flies, both in CH 6 dynamization and applied every 10 and 5 days, respectively, significantly reduced larval incidence compared to the control treatment.

Aiming to contain populations of *Thrips tabaci* (Thysanoptera: Thripidae) in organic potato crops, Gonçalves *et al.* (2009) noted that spraying *Calcarea carbonica* CH 6 and CH 30 decreased the incidence of thrips.

Rauber *et al.* (2007) evaluated the effects of agro-homeopathics on productivity and resistance of different potato genotypes, and confirmed that the preparation *Thuya* in potency CH 60 positively influenced the incidence of natural enemies of pests in trials on three potato genotypes.

Studies on the influence of agro-homeopathics and other natural insecticides on the natural enemies of insects of agricultural importance are especially necessary from the point of view of Integrated Pest Management (IPM), a phytosanitary control tool that associates different methods and applications. The conventional agricultural model, based on monoculture and the use of industrial inputs and defensives, has negatively affected the complex ecological dynamics, such as the balance between prey and predators, the richness and abundance of species and the selection of resistance. In view of this situation, IPM is a relevant instrument for the maintenance of organic farming systems, and research has already shown that the implementation of organic farming practices has a positive impact on the relationship between pests and their natural enemies, as well as on species diversity, richness and, above all, functional equity in communities (CROWDER *et al.*, 2010).

Krauss *et al.* (2011) compared 15 areas of organic triticale with 15 conventional areas, in order to evaluate the presence of pollinators, aphids and their predators. In the organic areas they observed twenty times more richness and one hundred times more abundance of pollinator species than in the conventional areas. In the organic areas the abundance of aphids was five times lower and the incidence of natural enemies was three times higher than in the conventional areas.

Jacobsen *et al.* (2019) investigated the abundance of the spider mite *Tetranychus urticae* (Acari: Tetranychidae) and its predators in conventional and organic strawberry crops, considering natural pesticide use and organic farming practices, such as crop edge vegetation patches.

The abundance of *T. urticae* in conventional crops was 10 times more than in organic crops, while the proportion of mites for their predators was 9.5 times less in organic crops. Therefore, the need for research examining the potential of agro-homeopathics for agricultural pest control, as well as their effects on non-target organisms such as natural enemies of pests, is warranted in order to assess their applicability in organic crops using multiple control agents within the IPM approach.

CASE STUDY WITH APHIDS

Among the pests that attack the cotton crop in Brazil, aphids have been highlighted. To control these insects, a large number of insecticide applications are generally used, promoting potential damage to the environment and to the farmer's health. As a result, interest in more sustainable technologies has increased. Considering that there are still few studies that focus on the potential of homeopathic insecticides and the impact of these products on the bioecology of pest insects, one of our works determined the lethal concentrations of Staphysagria CH6 in *Aphis gossypii* and the impact of the product on the reproduction of this aphid.

The research was conducted in the laboratory. After topical application of the homeopathic on the insects, daily evaluations were performed to determine survival. After the resurgence of the surviving insects, the aphids were individualized and daily reproduction was evaluated during the whole life cycle. The homeopathic impact on aphid population density was evaluated using an age structure model.

The lethal concentrations LC50, LC90 and LC99 were estimated at: 478, 66413, 133654 ppm. Negative effects of the lower concentrations of the homeopathic were found in the values of the life table parameters, such as liquid reproduction rate, doubling time and mean generation interval with respect to the control. In addition, there is evidence of sterilization effect of the surviving adults with the highest concentrations. Thus, the results obtained in this work are of great importance for subsidizing a sustainable management program for *A. gossypii*.

EFFECTS ON NATURAL ENEMIES

Through the dissemination of homeopathic preparations, which enable the natural control of pests, diseases and the balance of biological systems, we intend to propose practical means and instruments that contribute to the reduction of environmental impacts caused by the use of pesticides. Homeopathy does not harm the environment and can contribute to the sensitization of students to environmental issues, whether in the growth and/or development of plants, in the control of pests and diseases, improvement of crop productivity and in the natural defense of plants, as well as in all other segments of agriculture, livestock, veterinary and human medicine.

In order to analyze the impact of the alternative insecticide Staphisagria CH6 on the locomotion behavior of the ladybug *Harmonia axyridis*, a laboratory study was conducted.

The product was applied on cotton leaf discs. Adult females of the insect were kept together with one of the leaves in a Petri dish and were monitored with the aid of Ethovision equipment. Variables describing the insect's movement performance were explored in order to characterize the potential for behavioral alteration caused by Staphisagria CH6 in *H. axyridis*.

The results showed that under the influence of the agro-homeopathic product, the individuals presented alterations in their movement pattern, exhibiting higher average speed (cm/s) in the treatment with the product compared to the control. Despite this, the distance traveled (cm) remained higher in the control group (without application of the product). Thus, the results presented in this work are preliminary indications of the applicability of Staphisagria concomitantly with the use of *H. axyridis* as a biological control agent.

CONSIDERATIONS ON THE INSECTICIDAL ACTIVITY OF BIOACTIVE PLANTS

Plant extracts are substances from the secondary metabolism of plants, which are not essential for the generation of energy for the plant, such as substances from the primary metabolism, such as carbohydrates, lipids, amino acids and nucleotides. Secondary metabolites act as insecticides, feeding inhibitors and insect repellents (FAZOLIN *et al.* 2002).

Alkaloids, flavonoids, coumarins, tannins, quinones and essential oils are secondary compounds produced by plants for their survival (CASTRO, 2004). These compounds, after maceration, can be extracted in an aqueous medium or through organic solvents (WIESBROOK, 2004).

The secondary metabolites of plants can be divided into three chemically different groups: terpenes, phenolic compounds and nitrogenous compounds. Terpenoids have activities that suppress the development and appetite of insect larvae, in addition to acting as a repellent agent (JÚNIOR, 2003). While saponins are toxic to phytophagous in general (CAVALCANTE *et al.*, 2006).

Terpenes, pyrethroids and monoterpenes stand out as ingredients used in commercial insecticides due to low persistence, low mammalian toxicity and insecticidal activity (LINCOLN & ZEIGER, 2013).

Monoterpenes are found in trunks and branches of conifers, being toxic to a large number of insects, mainly to beetle species that are insect pests of these plants. Phenolic compounds act in defense against phytophagous and pathogens, among these compounds are tannins and flavonoids. Tannins can reduce the growth and survival of many phytophagous, acting as a food repellent for various animals. Flavonoids exhibit anti-feedant, sterilizing (by inhibiting ovarian development) and insecticidal properties (VENZON *et al.*, 2010). Research aimed at evaluating the insecticidal/acaricidal activity of plant extracts has been carried out to identify substances that affect insect behavior and metabolism.

The Brazilian flora is very rich in species with active principles of therapeutic importance, with potential not only for use in natural medicine but also in the integrated control of pests and diseases in agriculture, since many species of medicinal plants contain phenols, quinones, flavonoids and terpenoids in appreciable quantities (CARVALHO *et al.*, 2002). The insecticidal activity of essential oils can occur in several ways, causing mortality, deformations in different stages of development, as well as repellency and suppression, being that the repellent activity is the most common mode of action of essential oils and one of their main components (ISMAN, 2006).

The use of plant extracts represents a very important advantage due to their chemical complexity, since they can have beneficial synergistic effects, as opposed to what happens with a single active ingredient, which requires the use of very high doses and often causes adverse effects such as intoxication and environmental contamination. Another advantage is the presence of several metabolites in their composition, which hinders the evolution of resistance in insect populations (PERES, 2002).

Research carried out with insecticidal plants can have two important objectives: the discovery of molecules with action against insects that allow the synthesis of new insecticidal products and the obtaining of new natural insecticides for pest control. Examples include plants that have produced new synthetic products, such as *Physostigma venenosum* (Fabaceae), whose secondary compounds, especially physostigmine, served as a model for the synthesis of carbamates, and *Chrysanthemum cinerariaefolium*, which is the raw material from which pyrethrins, precursors of pyrethroids, are extracted (GALLO *et al.*, 2002).

Compounds derived from plants can act in different ways on insects, being repellents, oviposition inhibitors, feeding inhibitors, and can also alter the hormonal system, affecting development, causing deformations, mortality in the different stages and sterility (ROEL, 2001). According to Seffrin (2006), secondary metabolites are used by plants as a defense against insect attack, and the effects related to the selection behavior of insects with respect to the plant for feeding are among the most important.

Several researches have demonstrated the feasibility of using bioactive compounds obtained from plants for pest control, due to their efficiency, generally low cost, safety for applicators, consumers and environment (SHAAYA *et al.*, 1997, HUANG *et al.*, 2000, BOUDA *et al.*, 2001, DEMISSIE *et al.*, 2008). They can be used in the form of powders, aqueous or organic extracts, essential oils, emulsions, which present toxicity by contact, ingestion and/or fumigation (KARR & COATS, 1988, RAJENDRAN & SRIRANJINI 2008). These products promote mortality, repellency, impeded feeding and oviposition, and affect insect growth (HUANG *et al.,* 1999, MARTINEZ & VAN EMDEN 2001). The toxicity of essential oils on pests is influenced by their chemical composition, which depends on the plant resource, season of the year, ecological conditions, methods and time of extraction, and part of the plant used (LEE *et al.,* 2001).

Some plants produce secondary compounds that can be used for the development of new natural insecticides or be precursors for chemical semi-synthesis. Essential oil compounds are produced via the mevalonate metabolism pathway. Various substances such as phenylpropanoids and anethole in anise oil or orange oil have been associated with insecticidal activity (MORAIS, 2009). Some products formulated from vegetable oils (garlic, soybean, canola, cinnamon, neem, peppermint and cotton) are commercially available and have been recommended for the control of some species of aphids, mealybugs, whiteflies, thrips and mites (CLOYD *et al.,* 2009).

The compounds limonene, citronellol, citronellal, camphor and thymol were mentioned as repellents of various pests of the Orders Diptera, Coleoptera, Lepidoptera, Isoptera, Phtiraptera and Thysanoptera (NERIO *et al.,* 2010). The insecticidal compounds present in *Lonchocarpus* sp. are extracted from the root of the plant and are known as rotenoids. Rotenone initially causes a toxic effect on the muscles and nerves, rapidly ceasing insect feeding and causing their death a few hours or days after exposure. It is an insecticide and acaricide with a broad spectrum of action, and is used against beetle larvae, fleas, aphids, ants, cicadas, flies, mealybugs and mites. Rotenone is used extensively against the potato beetle *Leptinotarsa decemlineata* (Coleoptera: Chrysomelidae), which is one of the most important potato pests in the northern hemisphere (COSTA *et al.,* 1997; COX, 2002).

In trials conducted at the Phytotechnics Department of the Agricultural Sciences Center of the Federal University of Paraíba, Oliveira (2011) evaluated the potential use of natural products and found that anise oil caused high mortality (> 50%) of 1st, 2nd and 3rd instar larvae of Ceratitis *capitata* (Weid.) (Dipetera: Tephritidae), at concentrations higher than 1.5% (m/v). For proagrim, higher mortality was observed in 1st and 3rd instar larvae at concentrations of 2% and above, which was not confirmed in 2nd instar larvae. However, the mortality produced by orange oil was low in all concentrations. In the same study, it was found that the insecticidal activity of the products evaluated was not due to action via ingestion, since low mortality (< 50%) was confirmed in the trial in which the products were applied in the diet, regardless of the instar of the larvae of C. *capitata* evaluated. On the other hand, high mortality (> 50%) of the insect was observed in topical application trials, caused by products with anise and proagrim.

For the egg, 1st, 2nd and 3rd instar larvae and pupae stages of *C. capitata*, the mortality caused by the application of anise oil ranged from 23.75 (1%) to 33.70 (3%), 73.75% (1%) to 96.25% (3%), 78.75% (1%) to 97.5% (3%), 73.75% (1%) to 85% (3%) and from 7.7% (1%) to 12.50 (3%), respectively. It could be observed that the highest mortality rates were caused to the 1st, 2nd and 3rd instar larvae of *C. capitata* (OLIVEIRA *et al.*, 2011).

The products proagrim and anise oil caused high mortality in individuals of *C. capitata*, although mortality depended on the instar of the insect. For example, lower susceptibility of eggs and pupae of *C. capitata* to both products was recorded, probably due to the chemical constitution of the layer that covers the eggs and pupae of this pest, which hinders the penetration of the product (OLIVEIRA *et al.*, 2011).

The greater tolerance of eggs and pupae of *C. capitata* may possibly be related to the greater protection provided, in the case of the egg by the chorion, and greater rigidity of the tegument in pupae, offered by the envelope called puparium (KLOWDEN, 2009). In larvae, the penetration of these insecticides may have been facilitated through the greater contact provided by the routes of exposure (OLIVEIRA *et al.*, 2010), since these insecticides act by contact, both in the case of anise oil and/or as a fumigant in the case of proagrim. The substances present in these insecticides, such as phenylpropanoids and anethole (anise oil) (BARBOSA, 2009), triterpenoids and azadirachtin powder (proagrim), act through penetration into the insect's body via the respiratory system (fumigant effect) or through the cuticle (contact effect) in the case of proagrim (PRATES & SANTOS, 2000).

The insecticidal activity of orange oil observed in the study conducted by Oliveira (2011) was low in the immature stages of *C. capitata*, probably masked by the fact that this insecticide acts only by ingestion (digestive tract), which was not recorded in this work. According to Lopes *et al.* (2009), the activity of orange oil conferred by the substance STD (sodium tetraborohydrate decahydrate) caused high mortality of the aphid *Hyadaphis foeniculi* (Hemiptera: Aphididae), varying between 91.10% and 97.69% at concentrations of 0.3% and 0.7%, respectively. In general, the estimated LC50 of the Proagrim product for eggs and larvae of *C. capitata* and of anise oil for larvae of this insect represented a low volume to cause mortality to *C. capitata*.

Silva *et al.* (2010), from aqueous extract of neem almonds estimated LC50 for adults and immatures of *C. capitata* of 7,522 ppm (0.07522%) and 1,3668 ppm (1.3668%), and 13,028 (1.3028%) and 9,390 ppm (9.390%) for adults and immatures of *Anastrepha fraterculus* (Diptera: Tephritidae), respectively. Figueiredo *et al.* (2010) estimated low LC50 and LC90 of *Croton grewioides* (Baill.) (Euphorbiaceae) oil for *C. capitata* pupae, which were 0.29 and 0.95 % (m/v). However, the LC50 value observed by these authors is lower than that estimated in the present work, both for eggs and larvae in different instars. Moraes *et al.* (2006), believe that the insecticidal activity provided by plants of the genus *Croton* can be potentiated by a high concentration of substances such as anethole, methyleugenol, α-copaene, α-pinene, β-pinene and trans-caryophyllene.

Restello *et al.* (2009) recorded an average mortality of up to 94% when *Sitophilus zeamais* (Coleoptera: Curculionidae) was subjected to the application of *Tagetes patula* (L.) (Asteraceae) essential oil. Similarly, Lima *et al.* (2010) observed mortality higher than 70% of *Spodoptera frugiperda* (Lepidoptera: Noctuidae) from the concentration of 0.5% of the essential oil of leaves of *Ageratum conyzoides* (L.) (Asteraceae), these researchers also verified that at concentrations of 1, 2 and 3% (m/v) mortality was higher than 95%.

Pereira *et al.* (2008) used different types of essential oils in the control of *Callosobruchus maculatus* (Coleoptera: Bruchidae), and obtained results similar to those found in this study. These authors concluded that the essential oils of *Cymbopogon martini* Roxb. (Poaceae), *Piper aduncum* L. (Piperaceae) and *Lippia gracilis* Schauer (Verbenaceae) caused 100% mortality to *C. maculatus*. While the essential oil of *P. hispidinervum* caused 91.6% to 100% mortality, and the oil of *Melaleuca* sp. caused approximately 98% mortality to *C. maculatus*.

Due to the rapid evolution of resistance of the pest *Bemisia tabaci* Genn. (Hemiptera: Aleyrodidae) to insecticides (PRABHAKER *et al.*, 1998) and of the other problems caused by these products in the agroecosystem, some alternative methods have been studied for its control, including plant extracts from various botanical families with more emphasis on meliaceae (COUDRIET *et al.*, 1985; ASIATICO & ZOEBISCH, 1992; CUBILLO et *al.*, 1994; GÓMEZ et *al.*, 1997; NARDO et *al.*, 1997).

Oliveira *et al.* (1995) through laboratory experiments, proved that the extract derived from flowers of *Camellia sinensis* L. (Theaceae) is toxic to the insect pest *Sitophilus zeamais* Mots. (Coleoptera: Curculionidae) when applied directly to the insects with a hand sprayer. The extract of *Pinper niger* L. was more than 90% efficient in the control of *Sitotroga cerealella* Oliver (Lepidoptera: Gelechiidae), with a stable residual effect up to 90 days after its application.

According to Martinez (2002), other limonoids such as 14-epoxiazadiradione, meliantriol, gedunin, nimbidin, nimbinem, nimbin, melianone, azadiractol, vilosinin and melicarpine are also present in the neem fruit and participate in the process of ecdysis interruption, potentiating the insecticidal effect of neem oil.

Martinez (2002) observed the same effect on *Stomoxys calcitrans* L. (Diptera: Muscidae) (stable fly), which was unable to develop in manure treated with 0.5% neem oil sprays, probably because azadirachtin penetrated the cuticle of the insects and inhibited chitin synthesis, causing dehydration and death. There is little information on the ovicidal action of bioactive nim compounds on insects. Evaluations show that applications of high concentrations of plant extracts result in little or no ovicidal effect (SCHMUTTERER, 1988).

Neem products generally do not cause ovicidal effect, but their residual effect is often long enough to prevent the first ecdysis of larvae hatching from treated eggs (SCHMUTTERER, 1988). Since the methanolic extract of the kernel of the neem seed is toxic for larval feeding, it can be deduced that products derived from neem may be suitable and of great interest for the control of the tomato moth. In addition, according to Schmutterer (1990), it presents considerable selectivity for natural enemies of pests, especially parasitoids and predators, as well as it can be mixed with other bioproducts, such as insecticides based on *Bacillus thuringiensis* or as synergists to increase its effectiveness.

Substances of plant origin are used in the alternative control of *Zabrotes subfasciatus* in many countries of Latin America, Africa and Asia, in the form of extracts and oils, easily obtained and generally harmless to applicators and consumers. They cause mortality, repellency, inhibition of oviposition, reduction of larval development, fecundity and fertility of adults (WEAVER *et al.*, 1994). Several plant species were evaluated and showed considerable promise in the control of this pest, such as *Tagetes minuta* L. and *Ocimum canum* Sims (WEAVER *et al.*, 1994). *P. aduncum* (Piperaceae) is a plant of economic interest for the Amazon and can be used in pest control. This species produces an essential oil called dilapiol, whose insecticidal effect was described by Maia *et al.* (1988).

Several studies have shown that this plant, in addition to its medicinal importance as an anti-inflammatory, antihemorrhagic, astringent, diuretic and others, also presents insecticidal, bactericidal and fungicidal action (CORREA & PENNA, 1984; VIEIRA, 1991; VERAS, 2000; MORANDIM et *al.*, 2003; FIGUEIRA et *al.*, 2003; BASTOS et *al.*, 2003). According to Jacobson (1989), other botanical species of the Annonaceae, Asteraceae, Cannellaceae, Labiateae and Rutaceae families also seem to be promising for pest control. Studies on the insecticidal potential of most species are still scarce, which implies the need for the development of research for the discovery of new alternatives.

The main component of eucalyptus essential oil is 1,8-cinchol or eucalyptol. Its concentration is quite variable among eucalyptus species: *Eucalyptus citriodora* (Hook), *E. globulus*, *E. smithii* and *E. maidesii* (CHAGAS et al., 2002). The leaves of E. *citriodora* are repellent to the bean weevil A. *obtectus* (MAZZONETTO & VENDRAMIM, 2003) and possess insecticidal action on the beetles Tribolium *castaneum Herbst* (Coleoptera: Tenebrionidae) and Rhizopertha *dominica Fabr.* (Coleoptera: Bostrichidae). Pyrethrum is obtained from plants of the genus *Chrysanthemum*, of the Asteraceae family, which can be used to control insects such as aphids, larvae and chrysomelids. For the extraction of pyrethrum, the flowers of the plant are macerated. Its action can be increased with the use of sesame extract (*Sesamum indicum* L.) (GUERRA, 1985; SANTOS *et al.*, 1988).

Coumarin acts by irreversibly binding to cytochrome P450, compromising the insect's detoxifying capacity and inhibiting the electron transport chain, probably having as its site of action the cytochrome coxireductase (complex III), with a mechanism similar to that of β-methoxyacrylates (MOREIRA *et al.,* 2004). This botanical insecticide has action against larvae, beetles, ants, house flies (MOREIRA, 2002) and cockroaches. Sabadilla is extracted from the seeds of *Schoenocaulon officinale*, a member of the Liliaceae family. The main active compounds are alkaloids collectively known as veratrin, of which veratridine and cavadine have the highest insecticidal activity. Sabadilla can cause immediate mortality for some insect species, while others survive in a state of paralysis for several days before dying from loss of nerve function (CLOYD, 2004).

EFFECTS OF INSECTICIDAL PLANTS ON NATURAL ENEMIES

The use of insecticidal plants in the control of insect pests can be included in Integrated Pest Management Programs, since according to the FAO definition, IPM is a methodology that employs all economically, ecologically and toxicologically acceptable procedures to maintain populations of harmful organisms below economically acceptable levels, making the best possible use of the natural factors that limit the proliferation of these organisms.

Considering that the objective of IPM is to minimize the use of chemicals and prioritize biological, biotechnical and plant breeding measures, as well as cultivation techniques, the inclusion of control tactics that prioritize this objective has been increasingly implemented in this control program. The use of natural products as insecticides is compatible with the IPM philosophy, due to the fact that they have low toxicity to humans and other mammals, cause little interference to the environment, are selective and efficient against a wide range of insect pests. One of the main objectives with the use of plant products is the reduction of pest population growth, being that the collateral effects in the development that lead to the death of the insect is only one of the means of action of natural insecticides, since the use of high concentrations is necessary for this to happen (BOIÇA JUNIOR *et al.,* 2011).

Botanical extracts have some advantages over synthetic pesticides, such as: they offer new compounds that pests cannot yet inactivate, are less concentrated and potentially less toxic than pure compounds, and have rapid biodegradation and multiple modes of action, making broad spectrum use possible while retaining selective action within each pest class. At the same time, they are derived from renewable resources as opposed to synthetic materials (QUARLES, 1992).

Botanical insecticides ideal for use in integrated pest management programs should be toxic to pests, but not to natural enemies (PLAPP & BULL, 1978). Some botanical substances have known insecticidal action, such as: pyrethrins, rotenone, nicotine, cevadine, veratridine, ryanodine, quassinoids, azadirachtin and volatile biopesticides, the latter normally being essential oils present in aromatic plants (ISMAN, 2000). Being of natural origin, the insecticidal effect of the extracts tends to degrade after 72 hours, a positive characteristic because it leaves no residues and allows the use of other management techniques, such as the use of biological control, making the survival and performance of natural enemies viable, since it is more selective than chemical insecticides (MOREIRA *et al.*, 2006).

Silva *et al.* (2009) observed that anise essential oil (1%) reduced oviposition of *Euborellia annulipes* (Dermaptera: Anisolabididae), while aqueous tobacco extract (1%) was moderately selective for the postures of this predator. In addition, anise essential oil (1%) negatively affected the embryonic development of *E. annulipes*.

Costa *et al.* (2007) observed no influence of nim on the fertility of *E. annulipes* when their nymphs were previously treated with nim at concentrations of 0.5, 2.25 and 5%. In fact, the bioactive compounds of nim have been selective to non-target organisms, such as phytophagous mite predators compared to conventional acaricides (SPOLLEN & ISMAN, 1996; SCHMUTTERER, 1997).

Adults of *Podisus nigrispinus* (Heteroptera: Pentatomidae) were exposed to sprays of nim at different concentrations (0.0, 0.6, 0.8, 1.0, 1.5 and 2.0%) for 5 seconds. The results obtained showed that nim is selective for this insect at all concentrations evaluated, and that even the highest concentrations did not affect the reproductive performance of the predator (VACARI *et al.,* 2004). However, Cosme *et al.* (2007) studied the effect of botanical insecticides on the eggs and larvae of *Cycloneda sanguinea* (Coleoptera: Coccinelidae) and observed that egg viability was reduced in the case of azadirachtin application. Therefore, even if they are natural products, which are less aggressive compared to synthetic insecticides, the use of these products can have adverse effects on the beneficial insect population.

Torres (2004) studied the effects of the use of *Melia azedarach* and *Azadirachta indica* L. on the development of the parasitoid *Oomyzus sokolowskii* Kurdjunmov on cabbage cultivars and other plant species in the feeding of *Plutella xylostella* L. Thus, *M. azedarach* caused a reduction in the number of parasitoids and the rate of parasitism in the pupae of *P. xylostella*. In this study, it was also verified that the extracts of *M. azedarach* and *A. indica*, besides causing deformities in 10% of the adults of *O. sokolowskii* from the second generation, also caused a reduction in their size.

Mourão *et al.* (2004) studied the relative toxicity of *Acalypha indica* (neem) leaf, seed and oil extracts to the predatory mite *Iphiseiodes zuluagai* (Acari: Phytoseiidae) in the laboratory, and determined discriminatory concentrations (LCs99) of neem leaf, seed and oil extracts for adult females of *Oligonychus ilicis* (Acari: Tetranychidae). Through lethality concentration bioassays, they observed that the concentrations of the extracts of neem that killed 99% of *O. ilicis*, after 72 h of exposure were: 277.4; 520.9 and 10.9 mg/ml, of leaf, seed and oil, respectively. The discriminatory concentration of the oil extract of neem for adult females of *O. ilicis* was highly toxic to the mite *I. zuluagai*, and those of the leaf and seed extracts were highly selective.

Spraying 2% neem seed oil on *P. xylostela* eggs reduced the number of eggs parasitized by *Trichogramma principium* in the laboratory and by *T. pretiosum* in the field (KLEMM & SCHMUTTERER, 1993). RAGURAN & SINGH (1999) evaluated neem seed oil at concentrations of 5.0, 2.5, 1.2, 0.6 and 0.3% on oviposition, feeding, fecundity and development of Trichogramma *chilonis (Heteroptera*: Trichogrammatidae). The oil of neem seeds caused suspension of oviposition and feeding of the parasitoid.

Abrasom *et al.* (2006) stated that lavender oil has the advantage of attracting the ladybug *C. sanguinea*, a natural predator of aphids, and is less lethal than anise flowers. Bioactive neem compounds have shown relative selectivity to non-target organisms, such as predators of phytophagous mites, compared to conventional acaricides (MANSOUR *et al.*, 1987).

Cosme *et al.* (2007) observed prolongation of the last larval instar, morphological alterations and mortality of 1st and 4th instar larvae of *C. sanguinea*, as well as were also reported in treatments with Nim-I-Go at 50 and 100 mg L-1.

Souza (2010) observed 100% mortality of 2nd instar larvae of this ladybug when subjected to 0.0148 µg mL-1 a.i. of oil in the commercial formula DalNim. Venzon *et al.* (2007) showed lethal and sublethal effects on the ladybug *E. connexa* at concentrations of 0.25 and 0.5% of nim extract.

Most of the works on pest control with botanical products have emphasized the compatibility of these molecules with biological control. However, there are variations in the response of natural enemies to the application of such products (GONÇALVES-GERVÁSIO, 2003). The selective effect depends largely on the concentration of the active ingredient, as well as on the genetic, temporal and spatial variability of both the target pest population and the insecticidal plant.

EFFECTS OF ALTERNATIVE INSECTICIDES ON PARASITOID REPELLENCY

Natural enemies have the ability to identify a variety of potentially important chemicals that allow them to find several host species in the same area, as well as the ability to determine their value in the context of the environment, which may or may not be suitable for parasitism. Thus, the survival of progeny is directly related to this ability to correctly select the host during the search (TUMLINSON *et al.*, 1993). The two important sensory systems that control this behavior are olfactory and visual stimuli, which are perceived by peripheral receptors located on the antennae and compound eyes, and thus the behavior results in flight execution, host probing and oviposition (LEWIS & MARTIM, 1990).

Oliveira (2014) evaluated the effect of bioinsecticides applied on guava fruits, both repellency and attraction of *Diachasmimorpha longicaudata* Ashmead (Hymenoptera: Braconidae). Preliminary evaluations were carried out to select two varieties: one susceptible (Século XXI) and one resistant (Paluma). The bioinsecticide used was neem oil and water control, at the following conditions: 0.0; 3600; 5600; 10000 and 36000 ppm. The commercial product of neem oil used was AZAMAX®, a concentrated emulsion based on azadirachtin, of the tetranortriterpenoid group, with 1.2% azadirachtin (12 g/L). This product was introduced in the Brazilian market in 2009 and is the only one registered with the Ministry of Agriculture, Livestock and Supply (MAPA) for pest control in agriculture and certified by the Biodynamic Institute (IBD) for use in organic production systems.

In the host selection trial, the percentage of parasitism was not statistically different between the concentrations evaluated in any of the varieties studied. The same result was observed when the percentage of parasitism was compared between varieties, except in the control treatment, where it was verified that in the Século XXI variety the parasitism rate was higher (Oliveira, 2014).

Regarding larval mortality, no significant differences were found between the concentrations evaluated for both the Paluma and Século XXI varieties. Similarly, when mortality was compared between the varieties, except in the control, a higher mortality of larvae was found in the Paluma variety. However, in the joint evaluation of parasitism and larval mortality, no statistical difference was observed between the concentrations or between the varieties studied (Oliveira, 2014).

EFFECTS OF ALTERNATIVE INSECTICIDES ON PREDATOR REPELLENCY

In biological control programs with parasitoids, one of the difficulties in the release is related to the predation of parasitized eggs. Therefore, Mikami *et al.* (2014) selected predatory insect repellents that were selective to the parasitoid *Telenomus podisi*.

To examine repellency, Mikami *et al.*, (2014) made posters of *Euschistus heros* eggs and applied the essential oils. The posters were distributed in the soybean crop and removed after 24 h. Since cinnamon-leaf oil (100%) showed more repellency to predators (75 to 96.6%), it was applied at different concentrations on the parasitoid pupae and their emergence was evaluated. The concentrations used were: 0, 25, 50, 50, 75 and 100% oil, diluted in a 30% Tween solution. It was observed that the oil affected the emergence of the parasitoids in a concentration-dependent manner, even at the lowest concentration where it caused 30% mortality.

Alternatively, Mikami *et al.*, (2014) conducted evaluations with gelatin capsules for the utilization of cinnamon-leaf oil. Preliminary evaluations were performed, since there is no basic information on parasitoid release under these conditions. Evaluations were carried out with capsules with different colors, sizes and number of pupae/capsule. The size 0 capsule with 50% parasitized eggs showed the highest number of emerged insects, and color showed little influence on parasitoid emergence. Field evaluations were carried out to observe the protection of the capsules against predators. Thus, capsules with pupae were distributed in the soybean crop and collected after 24 h. The remaining pupae contained in the capsules were collected in the field. The remaining pupae contained in the capsules were evaluated, initial - final weight. The capsules did not show efficiency against predators, regardless of the colors used, and they are also extremely sensitive to humidity and therefore easily deformed. Therefore, the authors concluded that other alternatives should be studied to make the release of egg parasitoids feasible.

EFFECTS OF INSECTICIDAL PLANTS ON BEES

Bees are one of the most important groups of insects for humans because they allow the economic exploitation of their products and, mainly, because they contribute to the increase in the production of fruits and seeds of various plants. Approximately 90% of plant species have flowers and 75% of the world's agricultural crops need pollination, thus being considered the most important pollinators (Gianininni *et al.*, 2015).

Apis mellifera L. (Hymenoptera: Apidae) is of great economic importance, since it provides humans with bee products (honey, wax, royal jelly) and performs environmental services of ecological importance, such as pollination.

Beekeeping in Brazil began with the introduction of European *Apis mellifera* bees, but advanced with the help of African *Apis mellifera scutellata* Lepeletier in 1956, culminating with the beginning of the commercialization of honey, which is now marketed throughout the national territory. It is considered an alternative source of livelihood for the family farmer, since it improves the quality of life of the producers and does not cause damage to the environment (FREITAS *et al.*, 2004).

This activity contributes to the sustainability of small properties, providing direct or indirect employment opportunities in the management of bees, being more advantageous for small properties since it requires small areas, artisanal facilities and little labor to manage them (VARGAS, 2006).

Beekeeping is an activity that corresponds to the triangle of sustainability: social, economic and environmental. Social because it allows the fixation of human beings in the rural environment, economic because it is an alternative for the diversification of rural property and increased profitability, and constitutes an activity of great environmental importance, because it contributes to the maintenance and preservation of existing ecosystems (SILVA & PEIXE, 2012), being extremely important the preservation of A. mellifera bees.

Bees are excellent pollinators, frequently visiting crops that have been managed and received phytosanitary treatments. In that contact with the treatments, they can become contaminated and die, or then carry the contaminant back into the hive. Since small farmers and family farmers often have apiaries on their properties and generally use biological and/or alternative control for insect pest control in crops, it is important to evaluate the possible effect that these products may have on *A. mellifera* bees, a non-target organism.

The ideal phytosanitary product, from the point of view of agricultural production and Integrated Pest Management, is the one that presents total selectivity, that is, that offers lethal effect only on pests and preserves beneficial arthropods, avoiding ecological imbalance. The alternative control with the use of insecticidal plants is a set of methods that aims to offer alternatives to reduce the use of synthetic insecticides, also reducing the risks of contamination to the environment, human health, natural enemies and pollinators. Although several authors have analyzed the effect of biological and alternative products on pest insects, there is still little research with emphasis on their effects on useful insects, especially pollinators, such as *A. mellifera*.

It is extremely important that selective bioinsecticides are used for the preservation of beneficial species in the agroecosystem. Despite the importance of selectivity in the preservation of natural biological control of pests, there are still few studies on this subject. In this context, it is worth considering that the association between natural enemies and the use of plant extracts can constitute an alternative for integrated pest management. Although natural products are safer than synthetic insecticides, it is important to highlight their potential to have an effect on non-target organisms associated with crops, mainly pollinators, such as *A. mellifera* bees. Therefore, it is imperative to demonstrate the potential effects of plant extracts and/or alternative products before they are applied to non-target organisms, such as *A. mellifera bee* workers.

Advances in research involving alternative control are critical to aid in the control of insect pests, as well as in the preservation of beneficial insects and in the protection of the environment. One way to reduce the risks to pollinators is to conduct appropriate evaluations of botanical extracts in order to identify compounds that exhibit selectivity to bees. Selectivity can be understood as when the synthetic insecticide selects and/or controls the insect pest, without affecting the natural enemies. There are two types of selectivity: ecological selectivity and physiological selectivity. Ecological selectivity consists of the use of pesticide application techniques that minimize the contact between the product used and the non-target insect. While physiological selectivity seeks to identify compounds that are more toxic to pest insects than to beneficial insects (Pereira, 2016).

The use of alternative products favors the conservation of natural enemies, being adequate to use synthetic insecticides efficient in the control of insect pests and selective for natural enemies. Selectivity is one of the main characteristics for the choice of products to be applied, and the absence of this characteristic is one of the limitations of their use in alternative production systems.

In general, few studies have evaluated the selectivity of botanical extracts with homemade formulations in melipona and wild bees. What is known about bee tolerance to natural and synthetic toxins is that one of the main mechanisms used is metabolic resistance. And that the main enzymes responsible for the metabolism or detoxification of toxins are carboxylesterases, glutathione S-transferase (GSTs) and cytochrome P450 (Rand *et al.*, 2015). Still, there is a need for further work to understand the groups of chemical compounds present and the molecular weights of the extracts used to better elucidate the mechanisms that give these botanical extracts selectivity for bees. The mechanisms that allow bees to tolerate toxic secondary metabolites remain unknown.

In evaluating the effect of plant extracts on *A. mellifera* operans, Xavier (2009) verified that the product Rotenat® CE (a natural compound present in the plant *Derris* spp. (Leguminosae) caused 33% mortality after exposure to the product, but the same product was not toxic to the bee *Nonnotrigona testaceicornis* Lepetier (Hymenoptera: Apidae).

In the evaluation of the toxic effect of Rotenone (Rotenat® CE) on *A. mellifera* operans, Efrom (2009) found superior longevity of 48 hours for 69.5% of the bees sprayed with the product.

Silva et al. (2012) evaluated the effect of medicinal plant extracts on A. mellifera. In the ingestion experiment, it was demonstrated that among the seven treatments evaluated (powders of different plant structures [leaves and seeds]: neem [Azadirachta indica] 10%, neem seeds [A. indica] 7.9%, Mexican tea [Chenopodium ambrosioides L.] and ginger [Zingiber officinale L.] 60%), the treatment consisting of the Mexican tea and ginger extracts were more toxic, decreasing the survival of A. mellifera rapidly, by 65 and 60%, respectively.

When the action of the extracts was contact, the survival rates were close to the ingestion test in the case of Azamax® (85%) and Organic Nim® (85%), seeds (80%) and neem leaves (79%). The contact action was higher for Mexican tea (35%) and ginger (30%). Possibly, bee intoxication occurs by ingestion and contact. Thus, the concentrated extracts of C. ambrosioides L. and Z. officinale L. may act as phagoinhibitors or otherwise impair bee feeding, since they act as feeding interrupters (Santos et al., 2012).

Feeding cessation caused by such substances, due to reduced feed intake, results in nutritional deficiency. Lack of nutrients can cause developmental delay or deformities. Similarly, the occurrence of deformities or nutritional deficiency also reduces the insect's capacity for locomotion, the search for better quality food or places for shelter or reproduction, making it more susceptible to attack by natural enemies.

In another study conducted by Silva (2009), it was shown that ginger extracts did not affect the survival of A. mellifera. The treatment of neem leaves reduced the survival of bees by 25%, being higher than the rate found with extracts of neem seeds (20%), which contains the highest amount of the phytocomplex of this plant.

Pereira (2016) evaluated the selectivity of Nicotiana tabacum L. (leaf and dried leaf roll), Anadenanthera columbrina Vell. and Agave americana L. extracts on A. mellifera and Partamona helleri bees. In the contact bioassay, it was observed that N. tabacum (leaf) and A. colubrina extracts did not alter the survival of the two bee species compared to their respective controls, water with alcohol and water. N. tabacum extract (in roll form) decreased the survival of the two bee species evaluated. Contact with A. americana extract significantly reduced the probability of survival of A. mellifera, but did not modify the survival of P. helleri. A negative effect of the alcohol solvent used in the preparation of N. tabacum extracts (leaf and coil) on bee survival was evident. A. mellifera bees had lower survival probability in all treatments compared to the survival probability of P. helleri.

For the same trial mentioned above, even in the form of ingestion, Pereira (2016) detected that the survival of A. mellifera and P. helleri was affected by the extract of A. americana, while the extract of A. colubrina only affected the survival of A. mellifera. As noted in the contact tests, the alcohol present in the extracts also had a negative effect on bee survival. With the exception of the A. americana extract, A. mellifera bees were less likely to survive than P. helleri.

In general, the results obtained by Pereira (2016) revealed that the susceptibility of bees to the extracts varied between A. mellifera and P. helleri, according to the type of extract used in the bioassays and in the response to the type of exposure. After contact exposure, A. mellifera presented greater susceptibility to N. tabacum (coil) and A. americana extracts. In ingestion exposure, A. americana significantly reduced the survival of P. helleri and A. mellifera, and A. colubrina decreased the survival of A. mellifera. In general, the extracts were more selective for P. helleri.

In the same study carried out by Pereira (2016), it was revealed that after fasting A. mellifera ingested little contaminated food, regardless of the treatment, but when food free of extracts was provided, A. mellifera consumed more food. While P. helleri, after fasting ingested large amount of food and reduced the consumption of pure food over time, except in the N. tabacum roll (Pereira, 2016). Foraging melipona bees are able to perceive that they are at risk of enduring hunger and therefore, they carry a greater amount of food when they leave the hive to forage. Possibly, this strategy of A. mellifera explains the low consumption of contaminated food after fasting, and when provided with extract-free food they increase consumption to compensate for the low intake after starvation.

However, most records indicate that bees with larger body volume show more tolerance to pesticides, either by contact exposure or ingestion (Johansen et al., 1983) and some authors report that stingless bees (Meliponini) are more sensitive to pesticides (Tomé et al., 2012; Del Sarto et al. , 2014). The results showed that P. helleri was more tolerant to the extracts than A. mellifera. The lower susceptibility of P. helleri to extracts may be related to several aspects besides species and body size, such as, genetic differences, life cycle, feeding, foraging behavior (foraging) and type of exposure.

Other plant extracts were also evaluated on A. mellifera beekeepers, and different effects on these bees were found. Malerbo-Souza et al. (2003) found that when citronella (Cymbopogon winterianus) and oregano (Origanum vulgare) were sprayed at a concentration of 5% glycerin, water and oil on the passion fruit crop and/or in propylene tubes, there was no repellent effect on A. mellifera bees.

Vilani (2013) verified that Hovenia dulcis (raisin tree) extract reduced the longevity of A. mellifera by 54.20 h when sprayed (at a concentration of 1mL of plant extract at 5%) on A. mellifera adults. In studies that evaluated the selectivity of natural phytosanitary products, Xavier (2009) showed that Natualho®, at the recommended dose (1mL/200mL of solution) presented toxicity in adults of A. mellifera, causing 55% mortality after four days of exposure on pumpkin leaves immersed in aqueous solution containing botanical insecticides.

Efrom (2009) observed that when Natunim® (500 mL/100L) and Pironat® (250mL/100L) were applied via contact (in concentrations of 0.25×, 0.5×, 1× and 2× the recommended dose) on A. mellifera workers, they did not interfere with the longevity (48 h) of the bees. Simionatto (2013) when conducting studies spraying Natunim® (0.5mL/1L) directly on A. mellifera workers, verified that the product does not interfere in the longevity of the bees after 82 h.

Silva (2014) analyzed the effects of plant extracts of pomegranate (Punica granatum L.), spoonbill (Echinodorus grandiflorus), marjoram (Origanum majorana L.) and chamomile (Matricaria recutita L.). The study was carried out on A. mellifera operators, in four different ways of application: i) direct spraying of the treatments on the operator bees, ii) contact with the treatments on the sprayed glassy surface, iii) contact with the soybean leaves immersed in the solution of the treatments and iv) combination of the treatments in Candy cream. The results obtained by Silva (2014) provide important information for the optimization of bee conservation strategies in alternative insect pest management systems.

Marjoram is a plant from the Mediterranean region, known for its medicinal uses. The main constituents of marjoram oil are sabinene, alpha terpinene, gamma terpinene, cymene, terpinolene, linalool, sabinene hydrate, linalyl acetate, terpineol and gamma terpineol. The marjoram plant extract reduced the survival of A. mellifera operans in all bioassays performed. The marjoram extract caused morphometric modifications, decreasing the size of A. mellifera mesenteron cells, but did not cause morphological alterations (Silva, 2014).

As well as marjoram extract, pomegranate plant extract also promoted alterations in the size of mesentery cells of A. mellifera (Silva, 2014). Pomegranate is native to Asia and disseminated throughout the Mediterranean region, being cultivated almost everywhere in the world, including Brazil (LORENZI & MATOS, 2008). The main active components present in pomegranate are tannins (polyphenolic substances) and alkaloids, substances with antimicrobial action (PEREIRA et al., 2005). Pande & Akoh (2009) evaluated the antioxidant capacity and lipid profile of pomegranate, and found a higher concentration of hydrolyzable tannins in the peel. In general, antioxidant capacity was found in the leaves, followed by the peel, pulp and seeds. Gandhi et al. (2010) prepared pomegranate leaf powders to evaluate the insecticidal effect on the beetle Tribolium castaneum (Coleoptera: Tenebrionidae), and observed a high level of efficiency at a dose of 1 gram causing on average 82% mortality.

REPELLENT ACTIVITY OF ESSENTIAL OILS ON BEES

Despite the benefits provided by bees to humans, concern with accidents is associated with the frequency of swarms, which occur three to four times per year (DINIZ, 1990), and the variety of shelters in urban areas. Such shelters increase the contact between the insect and the human population. Direct contact situations usually occur when people inadvertently tamper with nearby areas or the premises where shelters are located, throw objects and apply chemicals, attempt to remove or destroy shelters without adequate protection, or some eventual contact with a single insect. Africanized bees are characterized by being very aggressive and attack their victims in swarms, inoculating large quantities of venom.

During the defense process of stinging bees, the operators inject the venom using the stinger, which remains inserted in the victim together with the viscera and the venom gland, ensuring a higher injected dose and increasing the efficiency of the defensive action. This process of loss of the stinger and attached parts results in the death of the insect, which only stings once. Individuals of species that do not have stingers or that have few serrated needles on their stingers may sting their victims more than once, since the stinger can be removed and inserted several times, as in wasps, for example. Considering the risks to the human population and the large number of accidents that could be avoided, it is extremely important that research be carried out to find repellent substances.

Essential oils occupy a prominent place in the agricultural defensives industry, since they present fungicidal and insecticidal properties against pests that cause prejudice to farmers, leading to losses in productivity and crop quality (SIMÕES et al, 2004). Fávero (2014) evaluated the repellent action of essential oils of rosemary (Rosmarinus officinalis), lemongrass (Cymbopogon citratus), thyme (Thymus vulgaris), cedar (Juniperus virginiana), clove (Syzygium aromaticum) and peppermint (Mentha piperita) on A. mellifera operans, in semi-field trials and aggressiveness trials.

Among the compounds evaluated in Fávero's study (2014), the essential oils of lemongrass, peppermint and clove showed the greatest repellent effect, almost completely inhibiting bee visits to the treated feeders. Cedarwood essential oil was the compound that showed the least repellent effect, while the other oils evaluated showed satisfactory repellency, which became less and less effective over time. In the a posteriori tests, lemongrass essential oil caused less bee aggressiveness compared to the control, which may confirm the repellent power of this compound. Thus, according to the results obtained by Fávero (2014), lemongrass has more potential for the development of repellent formulations for Africanized bees.

REFERENCES

ABRAMSON, C. I.; WANDERLEY, P. A.; WANDERLEY, M. J. A.; MINA, A. J. S.; SOUZA, O. B. Effect of essential oil from citronella and alfazema on fennel aphids Hyadaphis foeniculi Passerini (Hemiptera: Aphididae) and its predator Cycloneda sanguinea L. (Coleoptera: Coccinelidae). American Journal of Environmental Sciences, v. 3, p. 9-10, 2006.

ALMEIDA, A.A. de. Homeopathic preparations to control *Spodoptera frugiperda* (J.E. Smith, 1797) (Lepidoptera: Noctuidae) in corn. Dissertation (Master in Phytotechnology) Universidade Federal de Viçosa, Viçosa, MG. 54f. 2003.

ARAÚJO FILHO, R. **Introdução à pecuária ecológica**: a arte e a ciência de criar animais sem drogas ou venenos. Porto Alegre: São José, 2000. 136p.

ARENALES, M.C.; ROSSI, F. **Produção orgânica de carne bovina**. Viçosa: CPT, 2000. 158p.

BARBOSA NETO, R.M. **Bases of homeopathy. Campinas: Homeopathy League**, Medical Course of the State University of Campinas, 2006. 71 p. Available at:<http://www.fcm.unicamp.br/homeopatia/biblioteca/BASESDAHOMEOPATIA.pdf> Accessed on: August 5, 2013.

BARBOSA SILVA, A. Biological aspects and toxicity of plant products to *Euborellia annulipes*. **Thesis** (Doctorate in Agronomy) - Universidade Federal da Paraíba - UFPB, Areia - PB. 2009.

BASTOS, C. N.; SILVA, D. H. M. M.; MAIA, J. G. S. Atividade bactericida e composição de óleos essenciais de *Piper spp*. **Documentos IAC**, Campinas, 74p., 2003.

BOIÇA JUNIOR, A. L.; SILVA, A. J.; BOTTEGA, D. B.; RODRIGUES, N. E. L.; SOUZA, B. H. S.; PEIXOTO, M. L.; SOUZA, J. R. Plant resistance and the use of natural products as control tactics in integrated pest management. In: BUSOLI, A. C.; FRAGA, D. F.; SANTOS, L. C.; ALENCAR, J. R. D. C. C.; GRIGOLLI, J. F. J.; JANINI, J. C.; SOUZA, L. A.; VIANA, M. A.; FUNICHELO, M. **Tópicos em Entomologia agrícola - IV**. Jaboticabal-SP, p. 139-158, 2011.

BENEZ, S.M. **Manual de Homeopatia Veterinária**. São Paulo: Robe Editorial, 2002, 594p.

BITENCOURT, D.P.; BONATO, C.M. **Simple homeopathy applied to environmental education**. Paraná State Secretary of Education and Department of Biology - Laboratory of Homeopathy. Peabiru, 2008. 26p.

BONATO, C.M. Homeopathy in Plant Models. **Cultura Homeopática**, São Paulo, p.24-28, n.21, 2007.

BONATO, C.M. Homeopathy in agriculture. In: ENCONTRO BRASILEIRO DE HOMEOPATIA NA AGRICULTURA, 1., 2009, Campo Grande - MS. **Resumos...** 14p.

BETTI, L.; LAZZARATO, L; TREBBI, G., NANI, D. The potential and need for homeopathy research in horticulture and agriculture. In: **Proceeding of the International Conference "Improving the Success of Homeopathy: a Global Perspective"**. London, pp 64-68. 2006.

BOUDA, H.; TAPONDJOU, L. A.; FONTEM, D. A.; GUMEDZOE, M. Y. D. Effect of essential oils from leaves of *Ageratum co-nyzoides*, *Lantana camara* and *Chromolaena odorataon* the mortality of *Sitophilus zeamais* (Coleoptera: Curculionidae). **Journal of Stored Products Research**, v. 37, p. 103-109, 2001.

BRAZIL. Normative Instruction No. 7, of May 17, 1999. Provides on the norms for the production of organic plant and animal products. Diário Oficial da República Federal do Brasil, Brasília, v.99, n.94, p.11-14, 19 May 1999.

BRUNINI, C; SAMPAIO, C. **Matéria médica homeopática**. IBEHE. São Paulo: Mythos 2. 200p. 1993.

CÂMARA, F.L.A. Controlando pragas e doenças com homeopatia, na agricultura orgânica. **Horticultura brasileira**, Brasília, v.28, n. 2, p.16-21, 2010.

CARVALHO, R. A.; LACERDA, J. T.; OLIVEIRA, E. F.; SANTOS, E. S. **Medicinal plant extracts as a strategy for controlling fungal diseases of yam in the Northeast.** EMEPA-PB, João Pessoa, 10p., 2002.

CAVALCANTE, Giani M. MOREIRA, Albert F. C. VASCONCELOS, Simão D. Potencialidade inseticida de extratos aquosos de essências florestais sobre mosca-branca. **Pesquisa Agropecuaria Brasileira**, Brasília, v.41, n.1, p.9-14, jan. 2006.

CHAGAS, A. C. S.; PASSOS, W. M.; PRATES, H. T.; LEITE, R. C.; FURLONG, J. Efeito acaricida de óleos essenciais e concentrados emulsionáveis de *Eucalyptus spp* em *Boophilus microplus*. **Brasilian Journal of Veterinary Research and Animal Science**, v. 39, p. 247-253, 2002.

CLOYD, R. Natural indeed: Are natural insecticide safer and better then conventional insecticide? **Illinois Pesticide Review**, v. 17, p. 1-3, 2004.

CLOYD, R. A.; GALLE, C. L.; KEITII, S. R.; KALSCIIEUR, N. A.; KEMP, K.E. Effect os commercially available plant-derived essential oil products on arthropod pests. *Horticultural Entomology, v.* 102, p. 1567-1579, 2009.

CORREA, M. P.; PENNA, L. A. Dicionário das plantas úteis do Brasil e das exóticas cultivadas. **Instituto Brasileiro de Desenvolvimento Florestal**. Rio de Janeiro. 138 p. 1984.

COSTA, J. P. da; BELO, M.; BARBOSA, J. C. Effect of timbó species (*Derris spp*.: Fabaceae) on populations of *Musca domestica* L. **Annais** da **Sociedade Entomológica do Brasil**, v. 26, p. 163-168, 1997.

COX, C. Pyrehrins/Pyrethrum. **Journal of Pesticide Reform**, v. 22, p. 14-20, 2002.

COUDRIET, D. L; PRABHAKER, N.; MEYERDIRK, D. E. Sweetpotato whitefly (Homoptera: Alevrodidae): Effects of neem-seed extract on oviposition and immature stages. **Environmental Entomology**, v. 14, p. 776-779, 1985.

CROWDER, D.W; NORTHFIELD, T.D.; STRAND, M.R.; SNYDER, W.E. Organic agriculture promotes evenness and natural pest control. **Nature**, v.466, p 109-112, jul. 2010.

CUBILLO, D.; QUIJIJER, R.; LARRIVA, W.; CHACÓN, A.; HILJE, L. Evaluation of the repellency of various substances on the whitefly *Bemisia tabaci* (Homoptera: Aleyrodidae). **Integrated Pest Management**, v. 33, p. 26-28, 1994.

DAVIDSON, W.M. Insecticidal tests with oils and alkaloids of Larkspur (*Delphinium consolida*) and Stavesacre (*Delphinium staphisagria*). **Journal of Economic Entomology**, v. 22, n. 1, p. 226-234. 1929.

DEMISSIE, G.; TESHOME, A.; ABAKEMAL, D.; TADESSE, A. Cooking oils and "Triplex" in the control of *Sitophilus zeamais* Motschulsky (Coleoptera: Curculionidae). *Journal* of *Stored Products* Research, v. 44, p. 173-178, 2008.

DINIZ, N. M. **Estudo dos processos de enxameagem e de abandono de colônias de abelhas africanizadas em zonas rurais e urbanas** [Thesis]. Ribeirão Preto: Faculdade de Medicina de Ribeirão Preto; 1990.

EFROM, Caio F. S. **Rearing of *Anastrepha fraterculus* (Wied.) (Diptera: Tephritidae) in artificial diets and evaluation of phytosanitary products used in organic production system on this species and beneficial insects**. PhD Thesis in Phytotechnology, School of Agronomy, Federal University of Rio Grande do Sul, Porto Alegre. August, p.89, 2009.

FÁVERO, Rafael. Repellency study with several products of natural origin on Apis mellifera workers in semi-field. 2014.

FAZOLIN, M.; ESTRELA, J. L. V.; ARGOLO, V. M. Utilizacao de medicamentos homeopáticos no controle de *Cerotoma tingomariannus* Bechyné (Coleoptera, Chrysomelidae) em Rio branco, Acre. 2000.

FAZOLIN, Murilo; ESTRELA, Joelma L. V.; LIMA, Aldair P. de; ARGOLO, Valdirene M. 2002. Evaluation of plants with insecticide to control the cow-the-bean. Rio Branco: EMBRAPA-CPAFAC, 42p (Boletim de Pesquisa e Desenvolvimento, 37).

FHB - **Farmacopeia Homeopática Brasileira**. Brasília, 3.ed. 364p. 2011.

FIGUEIRA, G. M.; DUARTE, M. C. T.; SILVA, C. A. L.; DELARMELINA, C. Atividade antimicrobiana do extrato e do óleo essencial de *Piper spp* cultivadas na colecção de germoplamas do CPQBA - Unicamp, **Horticultura Brasileira**, Campinas, v. 21, n. 2, 403p., 2003.

FIGUEIREDO, W. R. S.; OLIVEIRA, F. Q. de.; OLIVEIRA, R. de.; BATISTA, J. L.; BRITO, C. H. Bioactivity of oil from *Croton grewioides* on the control of mediterranean fruit fly. Engenharia Ambiental, v. 7, n. 4, 2010.

FONSECA, M. F & WILKINSON, J. As Oportunidades e os Desafios da Agricultura Orgânica. p. 249 a 280. In: LIMA, D. M. de A.; WILKINSON, J. **Inovação nas Tradições da agricultura familiar**. Brasília: CNPq / Paralelo 15, 2002. 400p.

FREITAS, D. G. F.; KHAN, A. S.; SILVA, L. M. R. Nível tecnológico e rentabilidade de produção de mel de abelha (*Apis mellifera*) no Ceará Rev. Econ. Sociol. Rural vol.42 no.1 Brasília Jan./Mar. 2004.

GALLO, D.; NAKANO, O.; NETO, S. S.; CARVALHO, R. P. L.; BAPTISTA, G. S.; FILHO, E. B.; PARRA, J. R. P.; ZUCCHI, R. A.; ALVES, S. B.; VENDRAMIM, J. D.; MARCHINI, L. C.; LOPES, J. R. S.; OMOTO, C. Entomologia Agrícola. Piracicaba: FEALQ, 2002. 920p.

GANDHI, Nirjara; PILLAI, Sujatha; PATEL, Prabhudas. Efficacy of pulverized *Punica granatum* (Lythraceae) and *Murraya koenigii* (Rutaceae) leaves against stored grain pest *Tribolium castaneum* (Coleoptera: Tenebrionidae). **International Journal of Agriculture & Biology, v.** 12, p. 616-620, 2010.

GIESEL, A.; BOFF, M. I. C.; BOFF, P. Estudo comportamental da ant cortadeira *Acromyrmex spp.* submetida a preparados homeopáticos. **Revista Brasileira de Agroecologia**, v. 2, n. 2, p. 1259-1262, 2007.

GLIESSMAN, S. R. **Agroecology: the ecologys of sustainable food systems.** ^{2nd} ed. [*S.l.*]: CRC Press, Taylor & Francis Group, 384 p.

GOMEZ, P.; CUBILLO, D.; MORA, G. A.; HILJE, L. Evaluation of potential repellents of *Bemisia tabaci* I. Commercial products. **Integrated Pest Management**, v. 46, p. 9-16, 1997.

GONÇALVES, P. A. S.; BOFF, P.; BOFF, M. I. C. Homeopathic preparations of shell limestone in the management of Thrips *tabaci* Lind. thrips, and relationship with onion productivity in organic system. **Revista Brasileira de Agroecologia**, v. 4, n. 2, p. 228-230, 2009.

GUERRA, M. S. **Receituário caseiro: alternativas para o controle de pragas e doenças de plantas cultivadas e seus produtos.** Brasília, EMBRATER, 166p., 1985

HAMLY, E.C. **A arte de curar pela homeopatia.** 1. ed. São Paulo: Roca, 1982. 113p

HUANG, Y.; HO, S. H.; KINI, R. M. Bioactivities of safrole and isosafrole on *Sitophilus zeamais* (Coleoptera: Curculionidae) and *Tribolium castaneum* (Coleoptera: Tenebrionidae). *Journal of Stored Products* Research, v. 92, p. 676-683, 1999.

HUANG, Y.; LAM, S. L.; HO, S. H. Bioactivies ofessential oil from *Ellateria cardamomum* (L.) Maton. to *Sitophilus zeamais* Motschulsky and *Tribolium castaneum* (Herbst). ***Journal* of *Stored Products* Research**, v. 36, p. 107-117, 2000.

ISMAN, M. B. Botanical insecticides, deterrents, and repellents in modern agriculture and an increasingly regulated world. **Annual Review of Entomology**, n. 51, p. 45-66, 2006.

ISMAN, M. B. Plant essential oils for pest and disease management. **Crop Protection**, v. 19, p. 603-8, 2000.

ISMAN, M.B. Botanical insecticides, deterrents, and repellents in modern agriculture and an increasingly regulated world. **Annual Review of Entomology**, [*s.l.*], v. 51, p. 45-66, 2006.

JACOBSEN, S.K.; MORAES, G.J.; SØRENSEN, H.; SIGSGAARD, L. Organic cropping practice decreases pest abundance and positively influences predator-prey interactions. **Agriculture, Ecosystems and Environment**, v.272, p 1-9. 2019.

JACOBSON, M. Botanical pesticides: past, present and future. In: ANARSON, J.T.; PHILOCENE, B. J. R.; MORAND, P. (Eds.). Insticides of plant origin. Washington: **America Chemical Society**, p. 1-10, 1989.

JÄGER, T. *et al*. Use of homeopathic preparations in experimental studies with abiotically stressed plants. **Homeopathy**, v.100, p.275-287. 2011.

JOHANSEN C. A., D. F. MAYER, J. D. EVES AND C. W. KIOUS. 1983. Pesticides and bees. **Environmental Entomology**. Pages 1513-1518, https://doi.org/10.1093/ee/12.5.1513

JUNIOR, A.A SILVA. Essentia herba - Plantas bioativas. Florianópolis: Epagri, v.1. 2003. 441 p.

KARR, L. L.; COATS, J. R. Insecticidal properties of d-limonene. Journal of ***Pesticide Science***, v. 13, p. 287-289, 1988.

KRAUSS, J.; GALLENBERGER, I.; STEFFAN-DEWENTER, I. Decreased functional and biological pest control in conventional compared to organic crop fields. **PLoS ONE**, v.6, n.5, p.1-9. 2011.

LEE, S. E.; LEE, B. H.; SHOI, W. S.; PARK, B. S.; KIM, J. G.; CAMPBEL, B. C. Fumigant toxicity of volatile natural products from Korean spicies and medicinal plants towards the rice weevil, *Sitophilus oryzae* (L.). ***Pest Management Science***, v. 57, p. 548-553, 2001.

LEWIS, W. J.; MARTIN JR, W. R. Semiochemicals for use with parasitoids: status and future. Journal of Chemical Ecology, v. 16, p. 3067-3089, 1990.

LIMA, R. K.; CARDOSO, M. G.; MORAES, J. C.; ANDRADE, M. A.; MELO, B. A.; RODRIGUES, V. G. Chemical characterization and insecticidal activity of the essential oil of *Ageratum conyzoides* L. on the corn borer caterpillar *Spodoptera frugiperda* (Smith, 1797) (Lepidoptera: Noctuidae). **Bioscience Journals**, v. 26, n. 1, p. 1-5, 2010.

LOOS, R.A. Preparados homeopáticos visando o controle de podridão apical, traça e broca pequena do tomatoiro. Thesis (Doctorate in Phytotechnology) - Universidade Federal de Viçosa, Viçosa, MG. 98f. 2006.

LOPES, E. B.; BRITO, C. H. DE; BRITO, L. M. P.; ALBUQUERQUE, I. C.; BATISTA, J. L. Effect of orange oil on fennel aphid control. **Environmental Engineering**, v. 6, n. 2, p. 636-643, 2009.

LORENZI, Harri; MATOS, F.J.Abreu. **Plantas Medicinais no Brasil: Nativas e Exóticas**, 2 ed. São Paulo, 2008.

MAIRESSE, L. A. S. Avaliação da bioatividade de extratos de espécies vegetais, enquanto excipientes de aleloquímicos. **Thesis** (Doctorate in Agronomy). Santa Maria: UFSM, 2005, 329p.

MAJEWSKY, V.; ARLT, S.; SHAH, D.; SCHERR, C.; JAGER, T.; BETTI, L.; TREBBI, G.; BONAMIN, L.; KLOCKE, P.; BAUMGARTNER, S. Use of homeopathic preparations in experimental studies with healthy plants. **Homeopathy**, v. 98, n. 4, p. 228-243, 2009.

MALERBO-SOUZA, D. T; NOGUEIRA-COUTO R. H.; COUTO L. de A.; SOUZA, J. C. 2003. Attractants for *Apis mellifera* bees and pollination in orange (*Citrus sinensis* L. Osbeck, var. Pêra-Rio). Brazilian Journal of Veterinary Research and Animal Science. v. 40, pp. 272-278.

MAPELI, N. C. *et al*. Influence of homeopathic preparations on the immigration rate and colony growth of aphids (*Brevicoryne brassicae* (L.)) on cabbage plants. **Horticultura Brasileira**. Botucatu, v. 22, n. 2, p. 481-481. 2004.

MARTINEZ, S. S.; VAN EMDEN, H. F. Growth disruption, abnormalities and mortality of *Spodoptera littoralis* (Boisduval) (Lepidoptera: Noctuidae) caused by azadirachtin. **Neotropical Entomology**, v. 30, p. 113-124, 2001.

MARTINEZ, S. S. The neem, *Azadiractha indica* - **Nature, multiple uses, production**. IAPAR, Londrina. 142p.; 2002.

MAZZONETTO, F.; VENDRAMIM, J. D. Effect of powders of plant origin on *Acanthoscelides obtectus* (Say) (Coleoptera: Bruchidae) in stored beans. **Neotropical Entomology**, v. 32, p. 145-149, 2003.

MIKAMI, A.Y.; POMARI-FERNANDES, A.; BORTOLOTTO, O.C.; SILVA, G.V.; BUENO, A.D.F. Selection of essential oils repellent to predators and selective to the parasitoid Telenomus podisi (Hymenoptera: Platygastridae). In Embrapa Soja-Resumo em anais de congresso (ALICE). In: CONGRESSO BRASILEIRO DE ENTOMOLOGIA, 25., 2014, Goiânia. Entomologia integrada à sociedade para o desenvolvimento sustentável: annais.[Londrina]: SEB, 2014.

MORAES, S. M.; CAVALCANTI, E. S. B.; BERTINI, L. M.; OLIVEIRA, C. L. L.; RODRIGUES, J. R. B.; CARDOSO, J. H. L. Larvicidal activity of essential oils from brazilian Croton species against *Aedes aegypti*. **Journal of the American Mosquito Control Association**, v. 22, n. 1, p. 161-164, 2006

MORAIS, L. A. S. Óleos essenciais no controle fitossanitário. In: BETTIOL, W.; MORANDI, M. A. B. **Biocontrol of plant diseases**: use and perspective. Jaguariúna: Empraba Meio Ambiente, p. 139-152, 2009.

MORANDIM, A. A.; NAVICKIENE, H. M. D.; REGASINI, L. O.; CORDON, T.; FERRI, A. F. Constitution and antifungal activity of the essential oils from the leaves and stems of *Piper aduncum L.*, *P. arboreum* Aublet and *P. tuberculatum* Jaca and from the fruits of *P. aduncum L.* and *P. tuberculatum* Jacq. **Documentos - IAC**, Campinas, 74p., 2003.

MOREIRA, M. D. Isolation, identification and insecticidal activity of

chemicals of *Ageratum conyzoides*. **Dissertation** (Master in Entomology) - Viçosa, UFV, 60p., 2002.

MOREIRA, M. D.; PICANÇO, M. C.; BARBOSA, L. C. de A.; GUEDES, R. N. C.; SILVA, E. M. da. Toxicity of leaf extracts of *Ageratum conyzoides* to Lepidoptera pests of horticultural crops. **Biological Agriculture and Horticulture**, v. 2, p. 251-260, 2004.

MURRAY, B. **Biological terrain assessment**: Why animals get sick Disponível em: http://www.acresusa.com,>Acesso em: 13 August 2013.

NARDO, E. A. B. de; COSTA, A. S.; LOURENÇÃO, A. L. Melia azedarach extract as na antifeedant to *Bemisia tabaci* (Homoptera: Aleyrodidae). **Florida Entomologist**, v. 80, p. 92-94, 1997.

NERIO, L. S.; OLIVERO-VERBEL, J.; STASHENKO, E. Repellent activity of essential oils: a revie. **Bioresource Technology,** v. 101, p. 372-378, 2010.

OLIVEIRA, M. M.; GOLDFARB, A. C.; OLIVEIRA, E. C. S. Efeitos dos extratos etanólicos de *Piper sp* (piperácea) e *Camelia sinensis* sobre o inseto praga *Sitophilus zeamais* (Coléoptero: Curculionoidae). In: Reunião Anual da SBPC, 47, **São Luis: Anais...**, 478p., 1995.

OLIVEIRA, F. Q. de.; BATISTA, J. L.; MALAQUIAS, J. B.; ALMEIDA, D. M.; OLIVEIRA, R. de. Determination of the median lethal concentration (LC50) of mycoinsecticides for the control of *Ceratitis capitata* (Diptera: Tephritidae). **Revista Colombiana de Entomología**, v. 36, n. 2, p. 213-216, 2010.

OLIVEIRA, F. Q. Alternative technology in the control of *Ceratitis capitata* and its implication in the quality of *Spondias purpurea* fruits. **Dissertation** (Master in Environmental Science and Technology). Universidade Estadual da Paraíba - UEPB, Campina Grande, PB. 2011, 62p.

OLIVEIRA, F. Q. Associação de variedades de guiaba, bioinseticidas e o parasitóide *Diachasmimorpha longicaudata* no controle de *Anastrepha fraterculus*. **Thesis** (Doctorate in Agronomy - Plant Production) - College of Agricultural and Veterinary Sciences - Unesp, Jaboticabal Campus. 2014.

PANDE G.; AKOH C.C. Antioxidant capacity and lipid characterization of six Georgiagrown pomegranate cultivars. **Journal of Agricultural and Food Chemistry**, v.57, n.20, p.9427-9436, 2009.

PEREIRA, Jozinete V.; PEREIRA, Maria do S. V.; HIGINO, Jane S.; ALVES, Pollianna M.; ARAÚJO, Cristina R. F. Studies with the extract of *Punica granatum* L. (pomegranate): In Vitro Antimicrobial Effect and Clinical Evaluation of a Dentifrice on Dental Biofilm Microorganisms. **Revista Odonto Ciência** -Faculdade Odontologia/PUCRS, v. 20, n. 49, jul./set. 2005.

PEREIRA, A. C. R. L; OLIVEIRA, J. V. de; GONDIM JÚNIOR, M. G. C.; CÂMARA, C. A. G. da. Insecticidal activity of essential and fixed oils on *Callosobruchus maculatus* (FABR., 1775) (Coleoptera: Bruchidae) in caupi [*Vigna nguiculata* (L.) Walp.] grains. **Science and Agrotechnology**, v. 32, n. 3, p. 717-724, 2008.

PEREIRA, R. C. Selectivity of botanical extracts to bees *Partamona helleri* and *Apis mellifera* / Renata Cunha Pereira. - Viçosa, MG, 2016.

PERES, L. E. P. Secondary metabolism. **Rev. USP**, v.12, n.3, p. 5-32, 2002.

PRABHAKER, N.; TOSCANO, N. C.; HENNEBERRY, T. J. Evaluation of insecticide rotation and mixtures as resistance management strategies for *Bemisia argentifolii* (Homoptera: Aleyrodidae). **Journal of Economic Entomology**, v. 91, p. 820-826, 1998.

PRATES, H. T.; SANTOS, J. P. Óleos essenciais no controle de pragas de grãos armazenados, p. 443-461. In: LORINI, I.; MIIKE, L. H.; SCUSSEL, V. M. **Grain Storage**. Campinas: IBG, 1000 p. 2000.

PRIMAVESI, A. M. **Manejo Ecológico de Pragas e Doenças.** 2nd ed. São Paulo: Expressão Popular, 2016.

RAJENDRAN, S.; SRIRANJINI, V. 2 Plant products as fumigants for stored-product insect control. ***Journal of Stored Products* Research**, v. 44, p. 126-135, 2008.

RAUBER, L.P.; BOFF, M.I.; SILVA, Z.; FERREIRA, A.; BOFF, P. Management of potato diseases and pests using homeopathic preparations and genetic variability. **Revista Brasileira de Agroecologia**, v. 2, n. 2, p. 1008-1011, 2007.

REINHART. V.E. On the benefit-risk relation of homoopathic veterinary medicines. **Veterinary review**, n.48, p.778-796, 1993,

RESTELLO, R. M.; MENEGATT, C.; MOSSI, A. J. Effect of the essential oil of *Tagetes patula* L. (Asteraceae) on *Sitophilus zeamais* Motschulsky (Coleoptera, Curculionidae). **Revista Brasileira de Entomologia**, v. 53, n. 2, p. 304-307, 2009.

REZENDE, J.M.. **Homeopathy Primer**. Practical Instructions Generated by Farmers on the Use of Homeopathy in the Rural Environment. Organic Producers of the Vertente do Caparaó Region. Viçosa - MG, 2003, 35p.

ROEL, A. R. Utilizacao de plantas com propriedades inseticidas: uma contribuição para o Desenvolvimento Rural Sustentável. **International Journal of Local Development**, v. 2, p.43-50, 2001.

ROSSI, F.; AMBROSANO, E.J.; MELO, P.C.T.; GUIRADO, N.; MENDES, P.C.D. Experiências básicas de homeopatia em vegetais: Contribuição da pesquisa com vegetais para a consolidação da ciência homeopática. **Cultura Homeopática**, São Paulo, v.3, n.7, p.12-13, 2004.

ROSSI, F.; ARÉVALO, R.A.; AMBROSANO, E.J.; GUIRADO, N.; AMBROSANO, G.M.B.; MENDES, P.C.D.; MOTA, B.; ATZINGEN, E.M.M.V.; MENUZZO, M.M.; VARELLA, A.S. Application of homeopathic preparation in the control of tiririca in agroecological area. **Revista Brasileira de Agroecologia**, Porto Alegre, v.2, n.2, p.870-873, 2007a.

ROSSI, F.; AZEVEDO FILHO, J.A.; MELO, P.C.T.; AMBROSANO, E.J.; GUIRADO, N.; SCHAMMASS, E.A. Organic potato cultivation with application of homeopathic preparations. **Revista Brasileira de Agroecologia**, Porto Alegre, v.2, n.2, p.937-940, 2007b.

RUPP, L.C.D.; BOFF, M.I.C; BOFF, P.; GONÇALVES, P.A.S.; BOTTON, M. High dilution of *Staphysagria* and fruit fly biotherapic preparations to manage South American fruit fly, *Anastrepha fraterculus*, in organic peach orchards. **Biological Agriculture & Horticulture: An International Journal for Sustainable Production Systems**, v. 28 n.1, 41 - 48p. 2012.

SANTOS, J. H. R. dos; GADELHA, J. W. R.; CARVALHO, M. L.; PIMENTEL, J. V. F.; JÚLIO, P. V. M. R. **Controle alternativo de pragas e doenças.** Fortaleza, UFC, 216p., 1988.

SCOFIELD, A. Homeopathy and its potential role in agriculture, a critical review. **Biological Agriculture and Horticulture**, 1984.

SCHEMBRI, J. **Conheça a homeopatia**. 3.ed. Belo Horizonte: Rona, 1992. 268p.

SCHMUTTERER, H. Potential of azadirachtin-containing pesticides for integrated pest control in developing and industrialized countries. **Journal of Insect Physiology**, v. 34, p. 713-719, 1988.

SCHMUTTERER, H. Propreties and potencial of natural pesticides from the in tree, *Azadirachta indica*. **Annual Review of Entomology**, v. 35, p. 271-279, 1990.

SEFFRIN, R. C. A. S. Bioactivity of plant extracts on *Diabrotica speciosa* (Germar, 1824) (Coleoptera: Chrysomelidae). **Thesis** (Doctorate in Agronomy). Universidade Federal de Santa Maria. Centro de Ciências Rurais. Programa de Pós-Graduação em Agronomia, RS, 2006. 83p.

SHAAYA, E.; KOSTJUKOVSKI, M.; EILBERG, J.; SUKPRAKARN, C. Plant oils as fumigants and contact insecticides for the control of stored-product insects. *Journal* of *Stored Products* **Research**, v. 33, p. 7-15. 1997.

SILVA, M. A. Avaliação do potencial inseticida de *Azadirachta indica* (Meliaceae) visando ao controle de moscas-das-frutas (Díptera: Tephitidae). **Dissertation** (Master in Entomology). University of São Paulo - Luiz de Queiroz College of Agriculture USP/ESALQ, 2010.

SILVA, R. C. P. A.; PEIXE, B. C. S. Estudo da Cadeia Produtiva do Mel no Contexto da Apicultura Paranaense - uma Contribuição para a Identificação de Políticas Públicas Prioritárias. **Secretaria de Estado a Agricultura e do Abastecimento.** Curitiba: SEAB, 2012.

Silva, R. T. L. Effect of entomopathogens and plant extracts on Apis mellifera L. (Hymenoptera: Apidae). - Dois Vizinhos: Dissertation (Master's Degree) - Universidade Tecnológica Federal do Paraná. Programa de pós-graduação em Zootecnia. Dois Vizinhos, 2014.

SIMIONATTO, D. Action of control agents on *Apis mellifera* (Hymenoptera: Apidae), 2013, 37f. Course Conclusion Paper - Bachelor's Degree in Animal Science, Federal Technological University of Paraná. Dois Vizinhos, 2013.

SIMÕES, C.M.O.; SCHENKEL, E.P.; GOSMANN, G.; MELLO, J.C.P.; MENTZ, L.A.; PETROVICK, P.R. Farmacognosia: da planta ao medicamento. 5 ed. Porto Alegre, RS: Ed. da UFSC, 2004.

SOUZA, J.L; RESENDE, P. **Manual de horticultura orgânica**. 3.ed. Viçosa, MG: Aprenda Fácil, 2011. 843p.

SUJII, E.R.; VENZON, M.; MEDEIROS, M.A; PIRES, C.S.S; TOGNI, P.H.B. Práticas culturais no manejo de pragas na agricultura orgânica. *In*: VENZON, M.; JUNIOR, T.J.P; PALLINI, A. (coord.) **Controle alternativo de pragas e doenças na agricultura orgânica.** Viçosa: EPAMIG, 2010, p. 143-168.

TEIXEIRA, M.Z.; CARNEIRO, S.M.T.P.G. Effects of homeopathic high dillutions on plants: literature review. **Revista de Homeopatia**, v.80, n.3/4, p.104-120. 2017.

TORRES, A. L. Efeito associado de variedades de repolho *Brassica oleracea var.* capitata e estratos aqueosos de espécies vegetais na biologia de *Plutella xylostella* (L., 1758) e no parasitóide *Oomyzus sokolowskii* (Kurdjumov, 1912). **Thesis** (Doctorate in Agronomy) - College of Agricultural and Veterinary Sciences, Paulista State University, Jaboticabal-SP, 2004, 88p.

XAVIER, V. M. Impacto de Inseticidas Botânicos sobre *Apis mellifera, Nannotrigona testaceicornis* e *Tetragonisca angustula* (Hymenoptera: Apidae). Master's Dissertation. Viçosa: UFV. 2009. 43p. 2

VARGAS, T. **Avaliação da Qualidade do mel produzido na região dos Campos Gerais da Paraná. 2006.** 148f. Dissertation (Master) - Universidade Estadual de Ponta Grossa, Ponta Grossa.

VÉRAS, S. M.; YUYAMA, K.; Controle da vassoura-de-bruxa do cupuaçuzeiro por meio de extrato de *Piper aduncum* L. In: Congresso Brasileiro de Defensivos Agrícolas Naturais. **Abstracts**. Fortaleza, Brazil. 32p., 2000.

VENZON, Madelaine; JÚNIOR, Trazilbo, J. P.; PALLINI, Angelo. Alternative control of pests and diseases: Use of botanical insecticides in pest control. Viçosa, Epamig, 2010.

VIEIRA, L. S. **Manual de medicina popular: a fitoterapia da Amazônia**. Faculdade de Ciências Agrárias do Pará. Belém. 248p. 1991.

VILANI, Andreia. **Atividade de Produtos Fitossanitarios Naturais sobre** *Anticarsia gemmatalis* **Hübner (Lepidoptera: Noctuidae),** *Bacillus thuringiensis* **subsp. kurstaki e Seletividade** *Apis mellifera* **L. (Hymenoptera: Apidae).** 2013. 87 f. Dissertação Mestrado. Pato Branco: UTFPR. 2013. 83p.

WEAVER, D. K.; DUNKEL, F. V.; POTTER, R. C.; NTEZURUBANZA, L. Contract and fumigant efficacy of powdered and intact *Ocimum canum* Sims (Lamiales: Lamia ceae) a gainst *Zabrotes subfasciatus* (Bohemann) adults (Coleoptera Bruchidae), **Journal of Stored Products Research**, v. 30, p. 243-252, 1994.

WIESBROOK, M.L. Natural indeed: are natural insecticides safer and better than conventional insecticides? Illinois Pesticide Review, Urbana, v.17, n.3, p.1 3, 2004.

WYSS, E.; TAMM, L.; SIEBENWIRTH, J.; BAUMGARTNER, S. Homeopathic preparations to control the rosy apple aphid (*Dysaphis plantaginea* Pass.). **The Scientific World**, v. 10, p. 38-48, 2010.

Buy your books fast and straightforward online - at one of world's fastest growing online book stores! Environmentally sound due to Print-on-Demand technologies.

Buy your books online at
www.morebooks.shop

Kaufen Sie Ihre Bücher schnell und unkompliziert online – auf einer der am schnellsten wachsenden Buchhandelsplattformen weltweit! Dank Print-On-Demand umwelt- und ressourcenschonend produziert.

Bücher schneller online kaufen
www.morebooks.shop

KS OmniScriptum Publishing
Brivibas gatve 197
LV-1039 Riga, Latvia
Telefax: +371 686 204 55

info@omniscriptum.com
www.omniscriptum.com

Printed by Books on Demand GmbH, Norderstedt / Germany